Synthesis, Properties and Applications of Intermetallics, Ceramic and Cermet Coatings

Synthesis, Properties and Applications of Intermetallics, Ceramic and Cermet Coatings

Guest Editor
Cezary Senderowski

Basel • Beijing • Wuhan • Barcelona • Belgrade • Novi Sad • Cluj • Manchester

Guest Editor
Cezary Senderowski
Faculty of Mechanical and
Industrial Engineering
Warsaw University of
Technology
Warsaw
Poland

Editorial Office
MDPI AG
Grosspeteranlage 5
4052 Basel, Switzerland

This is a reprint of the Special Issue, published open access by the journal *Materials* (ISSN 1996-1944), freely accessible at: https://www.mdpi.com/journal/materials/special_issues/Intermetallics_Ceramic_Cermet_Coatings.

For citation purposes, cite each article independently as indicated on the article page online and as indicated below:

Lastname, A.A.; Lastname, B.B. Article Title. *Journal Name* **Year**, *Volume Number*, Page Range.

ISBN 978-3-7258-3289-7 (Hbk)
ISBN 978-3-7258-3290-3 (PDF)
https://doi.org/10.3390/books978-3-7258-3290-3

© 2025 by the authors. Articles in this book are Open Access and distributed under the Creative Commons Attribution (CC BY) license. The book as a whole is distributed by MDPI under the terms and conditions of the Creative Commons Attribution-NonCommercial-NoDerivs (CC BY-NC-ND) license (https://creativecommons.org/licenses/by-nc-nd/4.0/).

Contents

About the Editor vii

Cezary Senderowski
Synthesis, Properties and Applications of Intermetallics, Ceramic and Cermet Coatings
Reprinted from: *Materials* 2022, 15, 8408, https://doi.org/10.3390/ma15238408 1

Cezary Senderowski, Andrzej J. Panas, Bartosz Fikus, Dariusz Zasada, Mateusz Kopec and Kostyantyn V. Korytchenko
Effects of Heat and Momentum Gain Differentiation during Gas Detonation Spraying of FeAl Powder Particles into the Water
Reprinted from: *Materials* 2021, 14, 7443, https://doi.org/10.3390/ma14237443 5

Tomasz Chmielewski, Marcin Chmielewski, Anna Piątkowska, Agnieszka Grabias, Beata Skowrońska and Piotr Siwek
Phase Structure Evolution of the Fe-Al Arc-Sprayed Coating Stimulated by Annealing
Reprinted from: *Materials* 2021, 14, 3210, https://doi.org/10.3390/ma14123210 23

Cezary Senderowski, Nataliia Vigilianska, Oleksii Burlachenko, Oleksandr Grishchenko, Anatolii Murashov and Sergiy Stepanyuk
Effect of APS Spraying Parameters on the Microstructure Formation of Fe_3Al Intermetallics Coatings Using Mechanochemically Synthesized Nanocrystalline Fe-Al Powders
Reprinted from: *Materials* 2023, 16, 1669, https://doi.org/10.3390/ ma16041669 39

You Wang, Nan Deng, Zhenfeng Tong and Zhangjian Zhou
The Effect of Fe/Al Ratio and Substrate Hardness on Microstructure and Deposition Behavior of Cold-Sprayed Fe/Al Coatings
Reprinted from: *Materials* 2023, 16, 878, https://doi.org/10.3390/ma16020878 55

Jarosław Mizera, Bogusława Adamczyk-Cieślak, Piotr Maj, Paweł Wiśniewski, Marcin Drajewicz and Ryszard Sitek
Impact of an Aluminization Process on the Microstructure and Texture of Samples of Haynes 282 Nickel Alloy Produced Using the Direct Metal Laser Sintering (DMLS) Technique
Reprinted from: *Materials* 2023, 16, 5108, https://doi.org/10.3390/ma16145108 68

Dominik Kukla, Mateusz Kopec, Zbigniew L. Kowalewski, Denis J. Politis, Stanisław Jóźwiak and Cezary Senderowski
Thermal Barrier Stability and Wear Behavior of CVD Deposited Aluminide Coatings for MAR 247 Nickel Superalloy
Reprinted from: *Materials* 2020, 13, 3863, https://doi.org/10.3390/ma13173863 77

Dominik Kukla, Mateusz Kopec, Kehuan Wang, Cezary Senderowski and Zbigniew L. Kowalewski
Nondestructive Methodology for Identification of Local Discontinuities in Aluminide Layer-Coated MAR 247 during Its Fatigue Performance
Reprinted from: *Materials* 2021, 14, 3824, https://doi.org/10.3390/ma14143824 88

Ewelina Białek, Maksymilian Włodarski and Małgorzata Norek
Influence of Anodization Temperature on Geometrical and Optical Properties of Porous Anodic Alumina(PAA)-Based Photonic Structures
Reprinted from: *Materials* 2020, 13, 3185, https://doi.org/10.3390/ma13143185 101

Philipp Keuter, Anna L. Ravensburg, Marcus Hans, Soheil Karimi Aghda, Damian M. Holzapfel, Daniel Primetzhofer and Jochen M. Schneider
A Proposal for a Composite with Temperature-Independent Thermophysical Properties: HfV_2–HfV_2O_7
Reprinted from: *Materials* **2020**, *13*, 5021, https://doi.org/10.3390/ma13215021 . . . . **118**

Mateusz Kopec, Stanisław Jóźwiak and Zbigniew L. Kowalewski
A Novel Microstructural Evolution Model for Growth of Ultra-Fine Al_2O_3 Oxides from SiO_2 Silica Ceramic Decomposition during Self-Propagated High-Temperature Synthesis
Reprinted from: *Materials* **2020**, *13*, 2821, https://doi.org/10.3390/ma13122821 . . . . **135**

About the Editor

Cezary Senderowski

Cezary Senderowski graduated in 1990 with an M.Sc. in Airplanes and Helicopters from the Faculty of Mechanical Engineering, Military University of Technology, Warsaw, Poland. In 2002, he received his PhD in Materials Science and Engineering from the Military University of Technology, Warsaw. From 2004 to 2017, he was affiliated with the Department of Advanced Materials and Technology, Military University of Technology in Warsaw, where he was a Research Associate Professor. In 2010 and 2015, he participated in research during two scientific internships at the E.O. Paton Electric Welding Institute of the National Academy of Sciences of Ukraine, Kyiv (Department of Protective Coatings). From 2017 to 2022, he carried out research in the Department of Materials Technology and Machinery at the University of Warmia and Mazury, Olsztyn, in Poland. Since 2022, he has been affiliated with the Faculty of Mechanical and Industrial Engineering at the Warsaw University of Technology as the head of the Mechanics and Printing Institute. His research activities have encompassed such fields as surface engineering, powder metallurgy, quantitative characterization of microstructure in polycrystalline and nanocrystalline materials, thermodynamical characterization of coating materials, the relationship between structure and properties of composite nanomaterials, conventional and unconventional manufacturing technologies, especially thermal spray technologies (D-gun, HVOF, plasma and arc) and nanostructurization powders by mechano-chemistry combined with alloying ball milling.

Editorial

Synthesis, Properties and Applications of Intermetallics, Ceramic and Cermet Coatings

Cezary Senderowski

Mechanics and Printing Institute, Faculty of Production Engineering, Warsaw University of Technology, Narbutta 85, 02-524 Warsaw, Poland; cezary.senderowski@pw.edu.pl

Citation: Senderowski, C. Synthesis, Properties and Applications of Intermetallics, Ceramic and Cermet Coatings. *Materials* **2022**, *15*, 8408. https://doi.org/10.3390/ma15238408

Received: 23 October 2022
Accepted: 17 November 2022
Published: 25 November 2022

Publisher's Note: MDPI stays neutral with regard to jurisdictional claims in published maps and institutional affiliations.

Copyright: © 2022 by the author. Licensee MDPI, Basel, Switzerland. This article is an open access article distributed under the terms and conditions of the Creative Commons Attribution (CC BY) license (https://creativecommons.org/licenses/by/4.0/).

The production of intermetallic and ceramic protective coatings can be relatively simple, beneficial, and highly predictable. However, comprehensive possibilities for the synthesis and application of this type of coating are still limited by the technological conditions of the synthesis process and coexisting physical phenomena [1]. The physicochemical, thermophysical and structural properties of the coating materials frequently change in situ under synthesis conditions [2,3]. Furthermore, the possible formation of new phases with a nanocrystalline structure in relation to the feedstock powder material could be observed [4,5].

This Special Issue is focused on various conventional synthesis methods of different intermetallic, ceramic, and composite coatings obtained by thermal (D-gun and Arc) spraying [6,7], the CVD process [8–10], magnetron sputtering [11], anodization [12], and during the sintering of aluminum, iron and particulate mullite ceramic powders using self-propagated high-temperature synthesis (SHS) [13].

With regard to FeAl-type intermetallic materials, the main challenge is the brittleness of the FeAl (B2) phase due to their long-range ordered (LRO) crystal structure which limits the plastic deformation of the formed material. Therefore, a unique problem involves the analysis of the synthesis conditions' effect on the deformation and strengthening mechanisms of the nominally brittle intermetallic and ceramic alumina phases formed in situ during the powder particle synthesis processes at the supersonic jet flow in the D-gun spraying (DGS) process [3,6]. This description requires a deep understanding of both the elementary structural transformations of the FeAl powder particles sprayed in the D-gun process into the water and the grain structure in the synthesized intermetallic coating sprayed into the steel substrate [6,14]. On the other hand, the effects of synthesis conditions such as the impact of particle velocity, the temperature, and dynamic pressure of the gaseous stream on the chemical and phase, composition, crystallographic and morphological microtexture, size of the crystallites, and the state of the grain boundaries in the particles and obtained coatings, as well as the degree of superstructure disorder with the identification of nano/ultrafine grain and subgrain areas, dislocation, and antiphase domains, are discussed in detail [6,14].

It should be mentioned that during the thermophysical analysis of both the feedstock powder material and coatings, different phenomena were considered. These include the exchange of momentum and convective heat transfer, as well as the thermal effects of phase changes after the melting of powder particles, and additional analytical and numerical analysis [3,6,14]. It was found in [6] that the gas parameters and thermodynamic state of combustion products at the time of a detonation explosion of the gas mixture in the C–J plane are sufficient to melt FeAl powder particles up to 5 μm in diameter. However, under conditions of thermal forcing in the C–J plane, when a shock wave passes through the powder particles, the changes in particle velocity are much faster than the particle temperature changes caused by convective heating. Therefore, the main factor affecting the heating and melting degree of FeAl powder particles is the gaseous combustion products of detonation.

The thermal diffusivity of Fe40Al at.% powder particles with a diameter of up to 60 μm (mainly determined by the thermophysical properties of the FeAl phase) does not constitute any barrier to reaching the temperature in the particle volume in accordance with temperature changes on its surface. However, the powder particles with a diameter of 40 μm will not be able to absorb enough heat to melt them for the determined thermo-gas-kinetic parameters of the gas-detonation stream and the thermo-physical properties of the Fe40Al at.% phase [6]. In addition, such phenomena are favored by the formation of oxide films on the surface of FeAl powder particles. Such oxide films were found to be formed in the DGS process, during which the passage of the detonation wave front occurs within 10^{-7}–10^{-5}s, and during the expansion of gaseous detonation products, where partial melting and oxidation on the surface of particles occur.

The layers of Al_2O_3 oxides formed naturally in situ under DGS conditions on the surface of FeAl particles additionally constitute a diffusion barrier for heat transfer, typical for oxide ceramics. Thus, it was concluded that the presence of partially melted particles depends also on the heat transfer efficiency resulting mainly from the dynamics of the DGS process as well as the alumina contained in an inhomogeneous composite-like structure formed in situ during the D-gun process [14].

Detailed results presented in [6,14] show that under D-gun spraying conditions, an inhomogeneous composite-like structure of the FeAl coating is formed by unmelted or partially melted particles that underwent geometrical changes during the gas detonation spraying process due to strong plastic deformation.

An alternative way of forming the intermetallic phase-based coating structure from the FeAl system using the arc spraying method was presented by Chmielewski et al. [7]. During the simultaneous melting of two different electrode wires, aluminum and low-alloy steel (98.6 wt.% Fe), a coating structure was obtained in situ in the arc spraying process with the participation of FexAly intermetallic phases. In the second step, high-temperature heating in the range of 700–900 °C for 2 h was applied. During such annealing, the share of FeAl intermetallic phases with a clear ordering of the FeAl (B2) phase after annealing at 900 °C increased, which ensures greater phase homogeneity of the lamellar structure of the coating, with an evident decrease in microhardness [7].

Regarding FeAl-type intermetallics, one long-lasting challenge is to modify these hardly processable materials to improve their consolidation, e.g., by the formation of a ternary matrix also with the participation of dispersion precipitations of oxide ceramics strengthening the matrix. This was possible by using powder mixtures of 28 wt.%. Fe and 52 wt.% Al, also with the participation of 20 wt.% reinforcing mullite ceramics, sintered at high temperature [13]. The theoretical model and further experimental verification proposed by Kopeć et al. [13] enabled the formation of an intermetallic-based composite reinforced with alumina oxides. During the process, the rapid temperature increase, generated during SHS, led to the melting of Al-based metallic liquid. The metallic liquid infiltrated the porous SiO_2 ceramics, and silicon atoms were transited into the Fe-Al intermetallic matrix. Subsequently, the formation of a Fe–Al–Si ternary matrix and the synthesis of oxygen and aluminum occurred. Synthesis of both these elements resulted in the formation of new, fine Al_2O_3 precipitates in the volume of the matrix. A model for the growth of ultra-fine Al_2O_3 oxides from the decomposition of SiO_2 silica during the SHS process was experimentally verified through vacuum sintering, combined with high-energy milling of reinforcement particles. This method enabled us to manufacture an IMC/Al_2O_3 composite with a permanent connection between the matrix and reinforcement.

The HfV_2-HfV_2O_7 composite was also proposed as a material with potentially temperature-independent thermophysical properties that could be obtained by combining an anomalously growing thermoelastic HfV_2 solid with negative thermal expansion HfV_2O_7 [11]. Based on the novel design concept, the synthesis of the proposed HfV_2–HfV_2O_7 composite material was performed with two study pathways by (1) the annealing of magnetron-sputtered HfV_2 thin films in the air to form a HfV_2O_7 oxide scale on the thin film surface and (2) the magnetron sputtering of HfV_2O_7/HfV_2 bilayers [11]. Finally, the reduction in the HfV$_2$–HfV$_2$O$_7$ crystalline formation temperature from 550 °C, as obtained upon annealing, to 300 °C using reactive sputtering enables the synthesis of crystalline bilayered HfV$_2$–HfV$_2$O$_7$ to be achieved.

The need for new materials in the photonics industry is reflected in current trends and studies of porous anodic alumina (PAA) as a multifunctional porous ceramic coating prepared by the anodization of aluminum [12]. Intensified research activities on PAA material development for the photonic properties revealed the influence of temperature on the quality of the PAA-based distributed Bragg reflector (DBR) structure fabricated in the oxalic electrolyte in the temperature range of 5–30 °C. This work reported for the first time the production of PAA-based DBR with a good-quality PSB resonance in the mid-infrared (MIR) spectral region, which can extend the application of the PAA-based photonic structures up to the MIR spectral range [12].

Since the extreme performance conditions of modern aircraft engine turbines require the use of heat-resistant materials, MAR 247 nickel-based alloy is also in the scope of the analyzed materials [8,9]. The maximum operating temperature of contemporary nickel superalloys is 1100 °C, which is why it is necessary to use protective coatings on the hot parts of the aircraft engine turbines [15].

The authors of [8] show that the aluminide coatings of various thicknesses and microstructure uniformity deposited by the CVD process performed at different parameters effectively form a thermal barrier and bond coating interacting with the external environment in the air atmosphere at 1100 °C for 24 h under the thermal stability test conditions. The structure and physical-chemical properties, combined with dense and pore-free aluminide coatings obtained by optimized parameters of the CVD process at 1040 °C for 12 h in a protective hydrogen atmosphere, improved the mechanical response, thermal stability, wear resistance and exhibited good adhesion strength to MAR 247 nickel superalloy substrate.

With the increasing demands of the aircraft industry, the increasing application of nondestructive testing to determine the process-dependent properties of the material and to further reduce the amount of experimental work and minimize the manufacturing costs are observed. This is also accompanied by ever-increasing demands for the scientific description of applied measurement methodologies, such as, e.g., "Nondestructive Methodology for identification of local discontinuities in aluminide layer-coated MAR 247 during Its fatigue performance" [9].

In this paper, unconventionally, the fatigue performance of the aluminide layer-coated and as-received MAR 247 nickel superalloy with three different initial microstructures (fine grain, coarse grain and column-structured grain) was monitored using nondestructive, eddy current methods. The aluminide layers of 20 and 40 µm were obtained through the CVD process in the hydrogen protective atmosphere for 8 and 12 h at a temperature of 1040 °C and an internal pressure of 150 mbar. It was found that the elaborated methodology is an effective tool to monitor the degradation of the material. Furthermore, the applied assessment enabled us to localize the area with potential crack initiation and its propagation during 60,000 loading cycles [9]. This was mainly influenced by the initial microstructure of MAR 247 nickel superalloy and the thickness of the aluminide layer synthesized in the CVD process.

As in any field of research, collaboration between different researchers is the key to innovation. Such opportunities are provided by the exchange of experiences based on the research results presented in this Special Issue, "Synthesis, Properties and Applications of Intermetallics, Ceramic and Cermet Coatings", where publications can be submitted until 30 June 2023.

In this Special Issue, the conditions of synthesis using various methods affecting the structure and functional properties of intermetallic, ceramic and cermet coatings, as well as the analysis of phase transformations, thermophysical properties and other performance properties of produced coatings were studied in detail. Such analysis includes residual stress, adhesion, thermal stability, corrosion resistance and abrasive wear mechanisms, as well as analysis of the geometrical structure of the surface layer of the coatings together with the fractal characterized by using the root mean square (RMS), and other analysis in terms of the comprehensive use of the coatings.

The Editors give special thanks to the authors and the editorial team of *Materials* for the collaborative and peer-review process. We hope you enjoy reading this Special Issue and find new concepts for present and future research works.

Funding: The DGS research was founded by National Science Centre, Poland, Research Project No. 2015/19/B/ST8/02000.

Institutional Review Board Statement: Not applicable.

Informed Consent Statement: Not applicable.

Data Availability Statement: Not applicable.

Conflicts of Interest: The author declare no conflict of interest.

References

1. Deevi, S.C. Advanced intermetallic iron aluminide coatings for high-temperature applications. *Prog. Mater. Sci.* **2021**, *118*, 100769. [CrossRef]
2. Reddy, B.V.; Deevi, S.C. Thermophysical properties of FeAl (Fe-40 at. % Al). *Intermetallics* **2000**, *8*, 1369–1376. [CrossRef]
3. Panas, A.J.; Senderowski, C.; Fikus, B. Thermophysical properties of multiphase Fe-Al intermetallic-oxide ceramic coatings deposited by gas detonation spraying. *Thermochim. Acta* **2019**, *676*, 164–171. [CrossRef]
4. Wang, H.-T.; Li, C.J.; Yang, G.-J.; Li, C.X.; Zhang, Q.; Li, W.Y. Microstructural characterization of cold-sprayed nanostructured FeAl intermetallic compound coating and its ball-milled feedstock powders. *J. Therm. Spray Technol.* **2007**, *16*, 669–676. [CrossRef]
5. Grosdidier, T.; Ji, G.; Bernard, F.; Gaffet, E.; Munir, Z.; Launois, S. Synthesis of bulk FeAl nanostructured materials by HVOF spray forming and Spark Plasma Sintering. *Intermetallics* **2006**, *14*, 1208–1213. [CrossRef]
6. Senderowski, C.; Andrzej, J.P.; Fikus, B.; Zasada, D.; Kopec, M.; Korytchenko, K.V. Effects of heat and momentum gain differentiation during gas detonation spraying of FeAl powder particles into the water. *Materials* **2021**, *14*, 7443. [CrossRef] [PubMed]
7. Chmielewski, T.; Chmielewski, M.; Piątkowska, A.; Grabias, A.; Skowrońska, B.; Siwek, P. Phase structure evolution of the Fe-Al Arc-sprayed coating stimulated by annealing. *Materials* **2021**, *14*, 3210. [CrossRef] [PubMed]
8. Kukla, D.; Kopec, M.; Kowalewski, Z.L.; Politis, D.J.; Jóźwiak, S.; Senderowski, C. Thermal barrier stability and wear behavior of CVD deposited aluminide coatings for MAR 247 nickel superalloy. *Materials* **2020**, *13*, 3863. [CrossRef] [PubMed]
9. Kukla, D.; Kopec, M.; Wang, K.; Senderowski, C.; Kowalewski, Z.L. Nondestructive methodology for identification of local discontinuities in aluminide layer-coated MAR 247 during its fatigue performance. *Materials* **2021**, *14*, 3824. [CrossRef] [PubMed]
10. Kopec, M.; Kukla, D.; Yuan, X.; Rejmer, W.; Kowalewski, Z.L.; Senderowski, C. Aluminide thermal barrier coating for high temperature performance of MAR 247 nickel based superalloy. *Coatings* **2021**, *11*, 48. [CrossRef]
11. Keuter, P.; Ravensburg, A.L.; Hans, M.; Aghda, S.K.; Holzapfel, D.M.; Primetzhofer, D.; Schneider, J.M. A proposal for a composite with temperature-independent thermophysical properties: HfV_2–HfV_2O_7. *Materials* **2020**, *13*, 5021. [CrossRef]
12. Białek, E.; Włodarski, M.; Norek, M. Influence of anodization temperature on geometrical and optical properties of porous anodic alumina (PAA)-based photonic structures. *Materials* **2020**, *13*, 3185. [CrossRef] [PubMed]
13. Kopec, M.; Jóźwiak, S.; Kowalewski, Z.L. A novel microstructural evolution model for growth of ultra-fine Al_2O_3 oxides from SiO_2 silica ceramic decomposition during self-propagated high-temperature synthesis. *Materials* **2020**, *13*, 2821. [CrossRef]
14. Fikus, B.; Senderowski, C.; Panas, A.J. Modeling of dynamics and thermal history of Fe40Al intermetallic powder particles under gas detonation spraying using propane-air mixture. *J. Therm. Spray Technol.* **2019**, *28*, 346–358. [CrossRef]
15. Mori, T.; Kuroda, S.; Murakami, H.; Katanoda, H.; Sakamoto, Y.; Newman, S. Effects of initial oxidation on beta phase depletion and oxidation of CoNiCrAlY bond coatings fabricated by warm spray and HVOF processes. *Surf. Coat. Technol.* **2013**, *221*, 59–69. [CrossRef]

Article

Effects of Heat and Momentum Gain Differentiation during Gas Detonation Spraying of FeAl Powder Particles into the Water

Cezary Senderowski [1,*], Andrzej J. Panas [2], Bartosz Fikus [2], Dariusz Zasada [3], Mateusz Kopec [4,5] and Kostyantyn V. Korytchenko [6]

1. Department of Materials Technology and Machinery, University of Warmia and Mazury, 10-719 Olsztyn, Poland
2. Faculty of Mechatronics, Armament and Aerospace, Military University of Technology, 00-908 Warsaw, Poland; andrzej.panas@wat.edu.pl (A.J.P.); bartosz.fikus@wat.edu.pl (B.F.)
3. Faculty of Advanced Technologies and Chemistry, Military University of Technology, 00-908 Warsaw, Poland; dariusz.zasada@wat.edu.pl
4. Institute of Fundamental Technological Research, Polish Academy of Sciences, 02-106 Warsaw, Poland; mkopec@ippt.pan.pl
5. Department of Mechanical Engineering, Imperial College London, London SW7 2AZ, UK
6. Kharkiv Polytechnic Institute, National Technical University, 61-002 Kharkiv, Ukraine; korytchenko_kv@ukr.net
* Correspondence: cezary.senderowski@uwm.edu.pl

Abstract: In this paper, dynamic interactions between the FeAl particles and the gaseous detonation stream during supersonic D-gun spraying (DGS) conditions into the water are discussed in detail. Analytical and numerical models for the prediction of momentum and complex heat exchange, that includes radiative effects of heat transfer between the FeAl particle and the D-gun barrel wall and phase transformations due to melting and evaporation of the FeAl phase, are analyzed. Phase transformations identified during the DGS process impose the limit of FeAl grain size, which is required to maintain a solid state of aggregation during a collision with the substrate material. The identification of the characteristic time values for particle acceleration in the supersonic gas detonation flux, their convective heating and heat diffusion enable to assess the aggregation state of FeAl particles sprayed into water under certain DGS conditions.

Keywords: D-gun spraying; FeAl intermetallic powder; gas detonation flux; heat transfer; two-phase metallization stream; particle thermal history

1. Introduction

A unique feature of DGS technology is very high kinetic energy and thermal energy accumulated in the two-phase (gas and powder) metallization stream (that cause a large volumetric strain of the powder particles heated to very high temperatures), which changes non-monotonically with the dynamics of the detonation gas flow stream [1–6].

Parameters that affect a detonation wave include the gas velocity (D), its pressure (p), density (ρ) and temperature (T). These parameters can be considered within the frame of hydrodynamic theory and unidimensional stationary detonation (Zeldowicz–von Neumann–Doering (ZND) theory) [7–10]. According to this theory, the detonation wave consists of the shock wave (FU), the chemical reaction zone and the detonation products zone.

The shock wave is typically reduced to the (1-1) surface of high discontinuity in the pressure (p), density (ρ) and temperature (T) of the gas blend at the moment of detonation (Figure 1) [8].

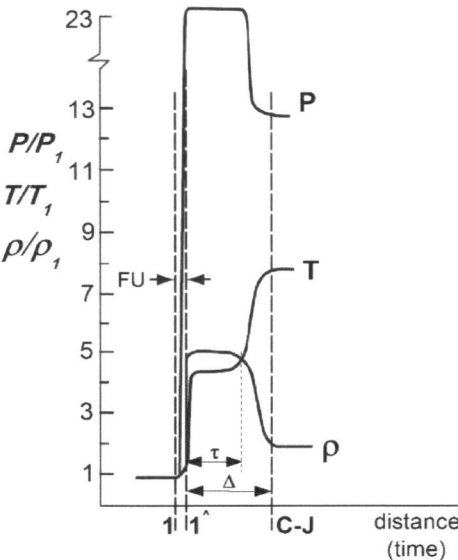

Figure 1. Change of gas thermodynamic parameters in the detonation wave area where: FU—shock wave, C-J—Chapman–Jouget plane.

The hypothetically infinitesimal chemical reaction zone is determined by the duration of the chemical reaction, that follows the wave and the velocity of detonation wave. Such velocity is constant and could be reached only after covering some distance (Δ) from the ignition point—in the Chapman–Jouget (C-J) plane.

Consequently, the system changes into the defined detonation wave, characterized by the detonation products, where temperature, density, pressure, velocity and chemical composition are dependent on the type of the gas blend.

Out of a series of interdependent conditions accompanying gas detonation, the composition of the explosive mixture has the greatest impact on the structure of the detonation wave, and thus on the value of thermodynamic parameters and the flow velocity of the two-phase metallization stream. It is mainly characterized by the type of working gas fuel and its share in relation to the oxidant (oxygen or air) and the method of diluting the explosive mixture by additionally introducing an inert gas as a diluter gas that delay the detonation. The analysis of the gas detonation mechanism and simulation data that is based on Euler's equations and ZND model, a strong relationship between the stoichiometry coefficient (λ) and the type of oxidant used with the detonation wave structure could be observed [2,4,9–12]. As already mentioned, this structure is determined by the thickness of the chemical reaction zone and the time of their course, which depends on the type of operating fuel.

The analysis of the relationship between the type and composition of the explosive mixtures and the detonation wave structure shows that the thickness of the chemical reaction zone and the time of its course reach the minimum values for approximately stoichiometric compositions regardless of the type of fuel and the accompanying oxidant [8]. The dimensions of the chemical reaction zone immediately downstream of the FU are always smaller for oxygen-fuel mixtures as compared to mixtures with air as the oxidant. The presence of nitrogen in the composition of the explosive detonation mixture with air (and also often intentionally introduced as a phlegmatizer to the fuel-oxygen mixture) limits the level of temperature and pressure that activate the processes of chemical reactions in detonation combustion after FU passing.

However, regardless of the type of demolition mixture used, the crossing of the detonation wave by the (1-1) surface results in the pressure increase (up to several atmospheres) followed by temperature increase (to a couple of thousand Kelvin above the critical temperature in which the gas enters the chemical reaction—Figure 1) [8,10–14]. Such phenomena are associated with a huge release of heat energy, with a simultaneous drop in pressure and density of the gas stream, which reaches the speed of sound at a Chapman–Jouget plane. Simultaneously, a multi expansion of the gaseous products, whose flow is being slowed due to the high pressure prevailing at the front of the moving concentrated shock wave, occurred (C-J plane—Figure 1). Such expansion results in rapid movement of the detonation wave toward the muzzle. As it reaches the open end of the barrel, the expansion of the gaseous detonation products at a supersonic speed with a series of successive stages of compression and expansion of gas products occurred [11,12].

The wave structure of the metallization stream's supersonic flow with the formation of Mach disks is relatively quickly extinguished by mutual friction of hot gas molecules with air molecules from the atmosphere induced by a turbulent shear layer of viscous damping [8]. However, damped oscillations of velocity and temperature of gaseous products, directly after their exit from the barrel of DGS (until the stream reaches the subsonic speed), also exert a significant influence on the final velocity and the degree of heating of the powder particles at the time of impact with the substrate material. This applies particularly to small particles with low inertia, which achieve the highest speed but are quickly slowed down [8,11,12].

In the final phase of the metallization stream flow, the thermal energy of the powder particles directly involved in the forming of the coating structure is determined by the dynamic pressure pulse at the moment of impact with the substrate, which induces significant changes in the temperature of the adhesion-diffusion junction. The literature shows, that for nickel powder feedstock with particle velocity greater than 500 m/s, a temperature increase results from the collision with the substrate. Such an increase of up to 300 °C [15,16] may in turn lead to the melting of the strongly softened powder particles.

Both the significant volume of the liquid phase (more than 10% vol.) resulting from the melting of the powder particles upon collision with the substrate during coating deposition, and the under-heating of the particles (powder temperature below 0.9 Tm—absolute melting point) result in a relatively high coating porosity of 4–6% [15]. The FeAl phase equilibrium system shows that the high-temperature FeAl solid solution (the composition of which is close to stochiometric) is stable at temperatures up to 1310 °C [8,17,18]. This fact is extremely important from a technological point of view. It can be hypothetically assumed that the lack of melting of the feedstock material during the DGS process enables the phase and chemical composition of the FeAl feedstock powder material to be maintained [19–22]. The fundamental problem raised during the optimization of the DGS process parameters is to ensure that the particles of the feedstock powder would retain the solid state while reaching their softening temperature. That enables the softened FeAl particles to undergo plastic deformation upon impact onto the substrate and geometrically transform them into the FeAl coating with unique properties [23–25]. Thus, the main aim of this work was to obtain data describing the process of heating and cooling of FeAl powder particles working under the dynamically changing conditions of the DGS spraying cycle. On the other hand, the work has been performed to develop and experimentally verify an analytical model of heat absorption by FeAl intermetallic powder particles in a flux of gaseous products during detonation combustion that occur during the D-gun spraying process of FeAl powder particles into the water. The analytical model of momentum and convective heat transfer between the FeAl particle and the gas flowing around them allowed for both, a precise description of such important phenomena during the gas detonation process, and a determination of the optimized conditions of FeAl particles thermal state. The model enables an estimation of the value of energy acquired by FeAl powder particles with different sizes in order to maintain the solid state in the D-gun spraying process. Thus, it is possible to determine the known size distribution of feedstock powder particles, the optimal parameters of DGS spraying at which FeAl powder particles reach the substrate in the solid state but highly softened.

On the contrary to known thermal spraying techniques [26], the detonation spraying procedure is characterized by a much higher complexity level [27,28]. It is because unstable gas flow conditions with high temperature and pressure gradients and high rates of local temporal DGS parameters change during the process (comp. e.g., [7,9–14]). During the analysis of the heat, momentum and mass transfer phenomena, the problems of coupling of various processes and their non-equilibrium character occurred. Therefore, such phenomena were subsequently confirmed by the assessment of heating degree of the FeAl powder particles under specific DGS spraying conditions into the water. It should be mentioned that the main work concentrated on demonstrating that for a given powder particle size distribution, only part of the particles would be melted. For this purpose, the parameters of the gas flow detonation were determined and the effects of heat and momentum exchange between the carrier flue gases and the powder particles were estimated.

2. Technological Conditions in DGS Process

Gas detonation spraying was performed at the E.O. Paton Electric Welding Institute of Ukrainian Academy of Sciences, Kiev by using the "Perun-S" detonation system (National Academy of Sciences of Ukraine, Kiev, Ukraine) (barrel length of 590 mm and 23 mm diameter).

An elementary working cycle of the DGS process starts with filling the channel with a specific explosive mixture. Subsequently, a certain portion (0.2–0.3 g) of the feedstock powder material is cyclically, axisymmetrically fed to the power injection position (PIP) by a carrier gas (i.e., nitrogen or air). PIP is the location of the powder in the barrel gun at the time of detonation. After such an injection, an electric spark with a constant frequency of 6.66 Hz is used to initiate detonation. Subsequently, the barrel line is rinsed with inert gas (nitrogen) after each unit cycle to prevent an excessive temperature increase of the barrel and substrate material, which during spraying does not exceed 100–150 °C.

In the presented research, the analysis of FeAl powder particles heating was compared to the unit operating cycle of the "Perun-S" gun during the actual DGS process of Fe40Al at.% powder particles into water, with the parameters presented in Table 1.

Table 1. DGS conditions for the 5–160 μm particles size of the Fe40Al at.% powder sprayed into water.

Detonation Gun Spraying Conditions	
Fuel gas (propane—C3H8)	0.45 m^3/h
Oxidant (oxygen)	1.52 m^3/h
Diluter gas (air)	0.25 m^3/h
Carrier gas (air)	0.4 m^3/h
Detonation frequency	6.66 Hz
PIP (powder injection position from ignition chamber)	275 mm
Spraying distance	110 mm

Commercial Fe40Al at.% powder with a wide particle size range from 5 μm to 160 μm (Figure 2), produced by using the VIGA method (vacuum inert gas atomization) in the CEA "LERMPS laboratory" in Grenoble, France, was used in this study. The IPS UA analyzer with an infrared diode (Kamika Instruments Company, Warsaw, Poland) was used to assess the particle size distribution of the FeAl powder particles. This analyzer enabled identification of the particle size with the additional possibility of precise counting in specific, declared size ranges. Scanning electron microscopy with X-ray microanalysis (Philips XL30/LaB6-DX4i-EDAX, Eindhoven, The Netherlands) was used to assess the degree of structural homogeneity (morphology and particle size of the FeAl powder in its as-received state and after DGS spraying into water).

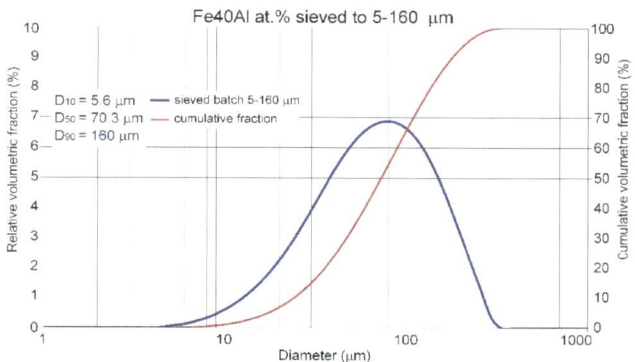

Figure 2. Particle size distribution of the Fe40Al at.% feedstock powder material as declared by the supplier.

Model Flow Parameters in DGS Process

The elementary work cycle of "Perun-S" lasts only 150 ms and it is repeated cyclically with a frequency of 6.66 Hz. In such a cycle a few important issues are considered:

- Thermodynamic properties and flow kinetics of gaseous detonation products, dependent mainly on the composition and the flow rate of the explosive mixture;
- Physicochemical and structural properties of the powder material, related to the average size, morphology, phase composition and the melting, evaporation and sublimation point of the powder particles.

Taking into account the frequency of shots of 6.66 Hz for the "Perun-S" gun during spraying FeAl coatings, the molar concentration of individual components of the gaseous explosive mixture at one working cycle of operation corresponds to the values presented in Table 2.

Table 2. Composition of detonation combustion products at DGS spraying of the Fe40Al at.% powder particles.

Working Explosive Gas Flow Rate [m³/h]: C_3H_8—0.45; O_2—1.52; Air—0.25						
Detonation Blend	Concentration			Fraction		(mol/1 Working Cycle in the DGS Process)
	(mol/h)	(g/h)	(% vol.)	(% mas.)	(% mol.)	
1	2	3	4	5	6	7
C_3H_8	20.08	883.34	20.30	26.20	20.30	0.84×10^{-3}
O_2	70.15	2444.76	70.80	66.50	70.80	2.93×10^{-3}
N_2	8.71	243.94	8.80	7.20	8.80	0.36×10^{-3}
Ar	0.10	4.16	0.10	0.10	0.10	0.43×10^{-5}

Calculation of the thermal parameters of the gas phase was performed using the TIGER 9.0 thermochemical code developed at Lawrence Livermore National Laboratory (LLNL) that considers the assumption of local chemical equilibrium for the detonation combustion conditions of a given propane explosive mixture with oxygen and air. Such code enables thermodynamic calculations for non-ideal, heterogeneous chemical systems of known atomic composition, containing gases and condensed phases (liquids and solid particles), described by the adopted gas state equations in detonation conditions [8]. Detonation parameters and the thermodynamic state of combustion products at the time of the detonation explosion of the mixture (in the Chapman–Jouget plane) were presented in Table 3. Weight and molar fractions of combustion products were calculated from the balance of chemical equilibrium using the libraries with free enthalpy values for specific reaction products implemented in the TIGER program (Table 4).

Table 3. Thermo-kinetic parameters of combustion products at the time of detonation in the Chapman–Jouget (C-J) plane determined using the thermodynamic code TIGER.

Parameter	Value
a (m/s)	1425
D (m/s)	2614
u (m/s)	1188
$\rho \left(\frac{kg}{m^3} \right)$	2.723
p (MPa)	4.444
T (K)	4461
$\kappa = \frac{c_p}{c_v}$	1.277
$\frac{\rho \cdot u^2}{2}$ (MPa)	1.92

a—speed of sound in the C-J plane. D—detonation velocity; u—mass velocity of detonation combustion products; ρ—density of gas detonation products; p—pressure; T—temperature; κ—isentropy exponent; $\frac{\rho \cdot u^2}{2}$—kinetic energy of the gas detonation products.

Table 4. Chemical composition of gaseous products determined using the TIGER thermodynamic code from the balance of chemical equilibrium of detonation combustion of a mixture with the composition presented in Table 2.

Product	Molar Fraction z_i, %	Mass Fraction g_i, %
H_2O	35.4997	29.8767
CO	33.6407	44.0411
H_2	15.3888	1.4390
CO_2	4.4557	9.3105
N_2	4.3309	5.6699
O_2	4.1090	6.1479
NO	2.5032	3.5111
NO_2	0.0017	0.0037
NH_3	0.0002	0.0002

In order to estimate the heat transfer parameters, it is necessary to determine both the thermo-kinetic properties of the explosive gas mixture under the applied detonation combustion conditions (presented above) and the thermo-physical properties of the FeAl type intermetallic powder material.

The characteristics of specific heat changes of the Fe40Al at.% feedstock powder material with the accompanying course of specific enthalpy changes as a function of temperature, were considered as presented in Table 5 and Figure 3 [8,14,29].

Table 5. Average results of the model calculated thermal-physical parameters for two characteristics temperature ranges corresponding to solid and liquid state of aggregation of the Fe40Al at.% powder material and the results of parameters of FeAl phase changes of melting and evaporation.

Parameter	Solid 0–1395 °C	Liquid 1395–2690 °C
Density ρ_d, kg/m³	5560	4806
Specific heat c_{pd}, J/(kg × K)	730	890
Thermal conductivity λ_{pd}, W/(m × K)	15	71
Thermal diffusivity a_d, m²/s	3.71×10^{-6}	16.6×10^{-6}
Surface emissivity ε_d	0.7	
Melting enthalpy Δh_{top}, kJ/kg	288	
Evaporation enthalpy Δh_{par}, kJ/kg	7364	

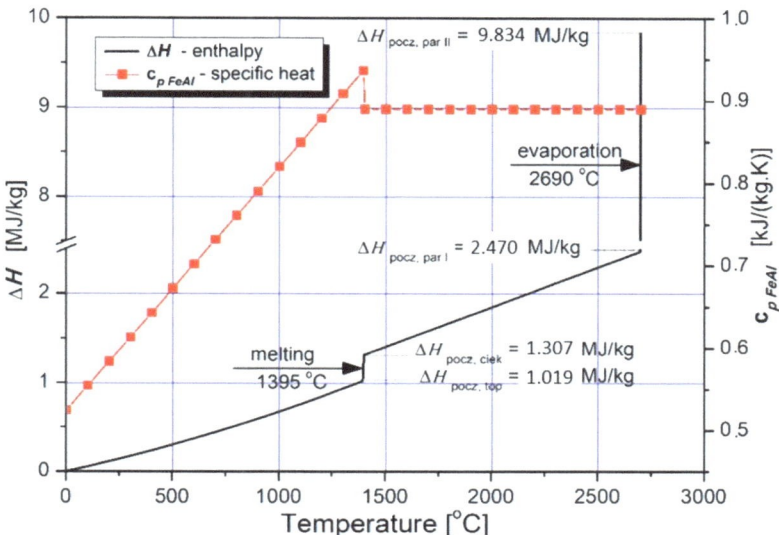

Figure 3. Dependence of the specific enthalpy of FeAl on the temperature (continuous line) obtained from model calculated dependence of the specific heat at the constant pressure with consideration of the value of enthalpy of phase changes of melting and evaporation (in accordance with Table 5).

Specific enthalpy is the main parameter that determines the amount of energy required for isobaric heating, melting and evaporating the powder feedstock material as a function of temperature. Despite the fact that the FeAl phase is stable up to 1310 °C (Figure 4), the first temperature range was selected to be from 0 °C to 1395 °C, which is the model liquidus temperature for the Fe40Al at.% alloy. The second range from 1395 °C to the conventional evaporation temperature of 2690 °C was determined on the basis of the arithmetic mean of the evaporation temperatures of Al and Fe [30–35].

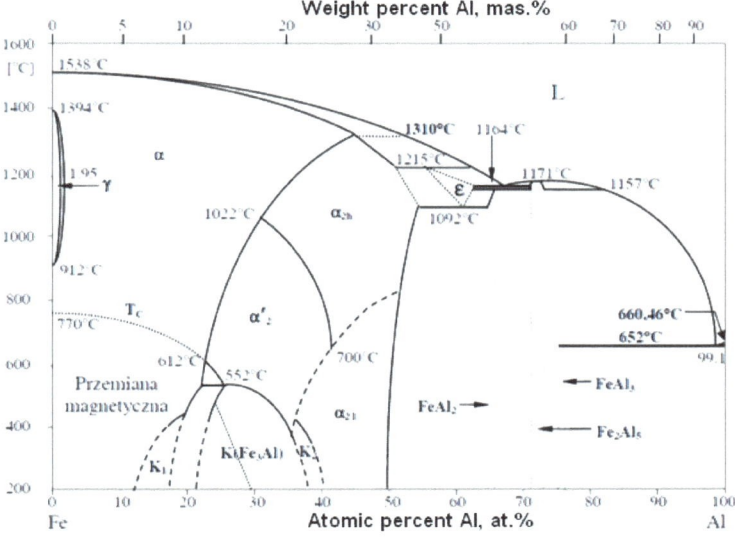

Figure 4. FeAl equilibrium system.

3. Assessment of Heat and Momentum Transfer Effects on FeAl Particle Thermodynamic State

The analysis of complex heat, momentum and mass transfer phenomena shaping the DGS spraying of FeAl particles was performed at two stages. First, through analytical considerations, it was proved that there is a certain particle diameter limit above which particles maintain their solid state [8,29]. Second, dynamics and thermal history of FeAl powder particles were analyzed through numerical calculations [14].

Details concerning modelling heat and momentum transfer between particle and detonation gases in a multiphase flow were provided in [8,14,29].

In general, the theoretical model for the momentum and convective heat exchange between the FeAl particle and the gaseous phase was considered to study effects such as particle melting, evaporation/sublimation, heat conduction in the volume of the FeAl particle and the radiation contribution to heat exchange between the FeAl particle and the walls of the detonation gun barrel. The analysis omitted the multi-phase structure of FeAl particles resulting from the presence of oxide phases formed in situ under DGS spraying conditions and the porosity of the powder particles. The sprayed FeAl particles were assumed as sphere shaped with an equivalent diameter, d. Moreover, the volume occupied by the powder particles was negligible in relation to the volume of the detonation gun barrel, so the presence of particles introduced in the amount of up to 0.25 g during the unit operation cycle of the gun, did not significantly affect the propagation of the detonation wave.

The dynamics of the flow and heat exchange process between the FeAl powder particle and the detonation wave, and the subsequent stream of gaseous products of the detonation combustion, make the particles heat up in a gas of variable velocity and with different temperatures. However, once the shock wave front (FD) has passed through, the parameters of the thermodynamic state of the gaseous detonation products in the vicinity of a single particle can be considered as homogeneous.

In the theoretical modelling, the ability of the powder particle to maintain a solid state at the moment of collision with the base material was based on the comparison of the modeled time values (τ_V, τ_T, and τ_a)—with respect to the approximate time of the FeAl particle exceeding the limit of the detonation combustion zone initiated by the shock wave (τ_A) and time of flight (τ_B)—from the point of insertion in the gun barrel (PIP) to the base material (water).

Where (τ_V, τ_T, and τ_a) mean respectively:

(τ_V)—particle acceleration time;
(τ_T)—particle convection heating time;
(τ_a)—temperature equalization time in the particle volume (heat diffusion).

The state of the FeAl particle was determined by comparing the enthalpy of the phase transition of melting and evaporation of the FeAl phase with the amount of energy (Figure 5) that the particle was able to absorb from the surrounding gas in the time from the transition of the detonation wave to the impact on the substrate. The main assumption involves the particle dwelling time in the interaction zone of gaseous products in the detonation combustion. The average velocity of FeAl powder particles, measured experimentally at the moment of their collision with the base material, for the technological conditions of DGS spraying in question, was approx. 730 m/s [8]. Thus, it was established that the total flight time of the powder particles to reach the "water level" in the tank, located 110 mm from the muzzle of the "Perun-S" gun, was 5.82×10^{-4} s, including the time of the particles remaining in the barrel of the "Perun-S" gun – 4.32×10^{-4} s that underestimates the real time of particle exposition to combustion gases.

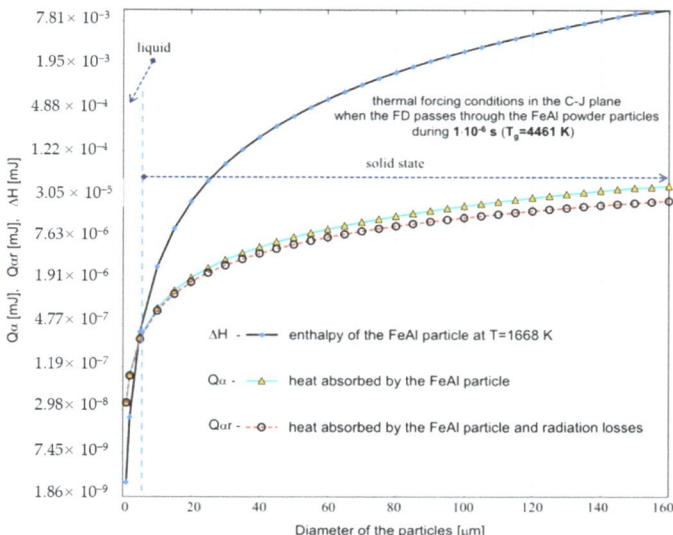

Figure 5. Comparison of the calculated values for the thermal energy Q_α absorbed by the powder particles with the enthalpy of the FeAl alloy in the solid state at the melting temperature ΔH, for different values of the particle diameter ranging from 1 μm to 160 μm. For the calculations, the parameters corresponding to the temperature of the gaseous medium of T = 4461 K during its interaction time of $\tau_A = 1 \times 10^{-6}$ s were adopted (the dashed line represents the results of the $Q_{\alpha r}$ calculations considering radiation losses).

Generally, the duration of the gas detonation process is very short and depends on the composition of the explosive mixture. According to the literature [2,9–14], the time of the detonation wave interaction in the C-J plane (τ_{C-J}) and the time of powder particle flight in the stream of gaseous products of detonation combustion (τ) are respectively: $\tau_{C-J} = 10^{-7} - 10^{-5}$ s and $\tau = 10^{-4} - 10^{-3}$ s.

Establishing specific velocities of the DGS process, the effectiveness of a specific method of heat exchange between the particle and the surrounding gas was assessed in terms of the heating of FeAl powder particles under the conditions of the DGS experiment (Figures 5 and 6).

Based on the analytical simplified modeling results, it can be concluded that particles with a diameter already above 40 μm will not be able to absorb enough heat to melt the FeAl material. This is evidenced by the analysis of the graphs shown in Figures 5 and 6—illustrating the comparison of the thermal energy Q absorbed by the particles of the powder charge with the enthalpy of the FeAl material in the solid state, at the melting temperature (ΔH), for the particle diameter ranging from 1 μm to 160 μm.

Whereby, under the thermal forcing conditions generated in the CJ plane, when the FD passes through the powder particles, the amount of thermal energy absorbed by the particle is limited by the short residence time of the particle in the detonation wave zone, and thus particles up to 5 μm in diameter are melted (Figure 5). In this case, the greater influence of radiation effects can be observed during the heat transfer between the FeAl particle and the walls of the detonation gun.

Under the thermal forcing conditions generated in the C-J plane, when FD passes through the powder particles, the particle velocity changes are much faster than the particle temperature changes caused by its convective heating. Thus, despite the negligibly low temperature equalization resistances in the FeAl particle (i.e., low heat conduction resistances), it will not obtain the gas temperature in its entire volume during the passage of the detonation wave, because the enthalpy of the particle increases by convective heat supply through its surface. Under these conditions, the amount of thermal energy absorbed by the particle is limited by the short residence time of the particle in the detonation wave zone.

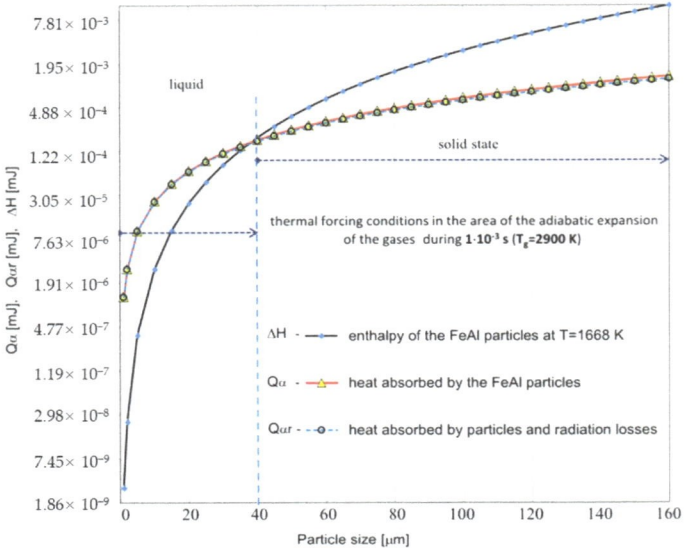

Figure 6. Comparison of the calculated values for the thermal energy Q_α absorbed by the powder particles with the enthalpy of the FeAl alloy in the solid state at the melting temperature ΔH, for different values of the particle diameter ranging from 1 μm to 160 μm. For the calculations, the parameters corresponding to the temperature of the gaseous medium of T = 2900 K during its interaction time of $\tau_A = 1 \times 10^{-3}$ s were adopted (the dashed line represents the results of the $Q_{\alpha r}$ calculations considering radiation losses).

As the powder particles accelerate the fastest when passing through the FD and then strive to equalize their velocity with the stream of gaseous products of detonation combustion, their main heating takes place in a slower flow of the gas stream (approx. 10−3 s) with lower values of the heat transfer coefficient. During the flow in the expansion stage of the gaseous products, the particle velocity gradually equalizes with the gas velocity. Due to the polydispersity of FeAl powder particles in the particle size range of 1–160 μm, they are characterized by a different flow velocity, spreading in the stream of hot gases, along the length of the barrel gun. As the biphasic metallization stream spreads, the amount of heat transferred to the finer FeAl particles remaining in the "tail of the stream" increases due to their lower flight speed and longer residence time in the gas stream. These particles, due to their greater inertia, experience less acceleration in the first phase of flight, when the detonation wave passes through them. Moreover, as the particle velocity equalizes with the gas velocity, the heat transfer coefficient decreases. At the same time, the corresponding convective heating time lengthens.

Hence, with the adopted assumptions model, the exact determination of the limit diameter of non-remelting particles is a complex issue. However, it is undisputed that such a boarder limit exists and further depends on the time that the particles spent in gaseous stream of the detonation products.

In order to verify the results and to further obtain a more precise estimation, the numerical calculations were performed using the Ansys Fluent CFD 2019 software, as described in [14].

Similar to the presented analytical modelling, the whole analysis was restricted to one single cycle of gas detonation and particle propulsion. Because the crucial stage of propulsion takes place in the interval of gas outflow from the barrel, therefore the numerical simulation was terminated when the largest particle reached the base surface, i.e., water level. For the momentum transfer and heat exchange between the gas phase and the powder particles, the same formulas as in the previous analysis were applied. However, the combustion reactions were modelled independently in order to match the parameters obtained from the TIGER code in C-J plane (as shown in Table 3). As a result of the conducted simulations, the spatial distributions of the gas physical parameters, as well as the time history of the particles' state and motion, were obtained. Comparison of the effects of theoretical modeling with numerical experiments suggests an acceptable discrepancy of the results. The comparison of theoretical modeling with numerical experiments indicates an acceptable discrepancy in velocity results of 920 and 710 m/s, for 10 and 20 μm diameter particles, respectively [14], which is consistent with the velocity measurement in the DGS experiment. In order to demonstrate the effect of heating during the whole DGS process, both the temperature changes in the surface of FeAl particles and the radial temperature distribution were obtained from the numerical analysis. The results were subsequently related to the heat transfer coefficient evolution as a function of time, which is extremely high for small particles exposed to high temperature, while undergoing the highest acceleration in the first phase, during the detonation wave interaction.

It was concluded that the thermal diffusivity of FeAl powder particles up to 60 μm in diameter (mainly determined by the thermophysical properties of the FeAl phase) did not constitute any barrier to reaching the temperature in the particle volume in line with temperature changes on its surface. Thus, basically the FeAl powder particles heat up in the gas stream evenly throughout their entire volume, regardless of their size.

In the course of numerical calculations, the critical diameter for unmelted particles was established to be equal to about 80 μm [14]. Collective results of the numerical analyses are presented in Figures 7 and 8.

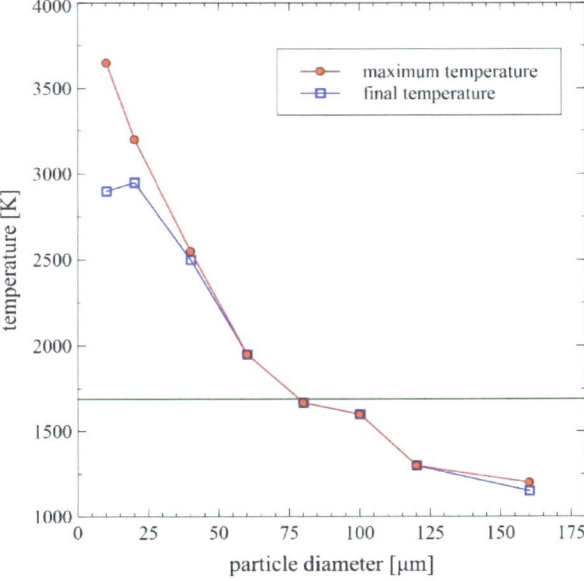

Figure 7. Comparison of the maximum and final particle temperature during the DGS spraying process as a function of the particle diameter—results of the DGS multiphase flow numerical modeling [14] with indication of the temperature of FeAl alloy melting.

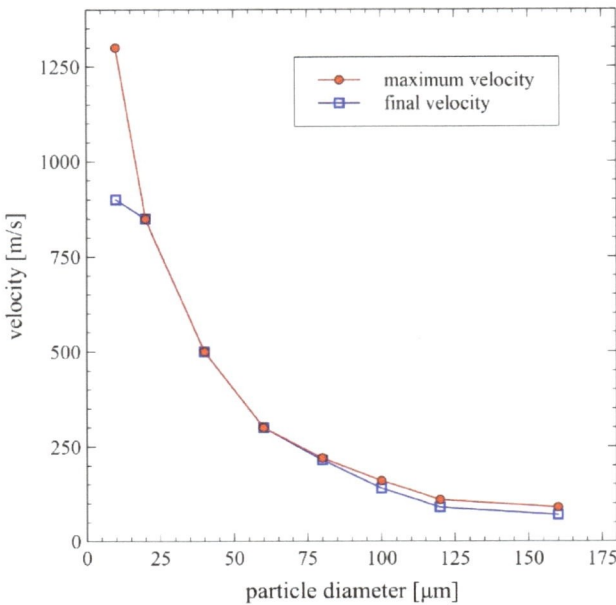

Figure 8. Maximum and final (at the base surface, i.e., water level) particle velocity as a function of its diameter—results of the DGS multiphase flow numerical modeling [14].

The analysis of the presented data confirms the exceptionally large range of the thermodynamic state of the sprayed particles, which is a distinguishing feature of the gas detonation spraying process. On the one hand, it presents a great difficulty in planning an experiment and developing a DGS technology. On the other hand, it could be used to specifically shape the properties of the coating. It should be highlighted, however, that this is certainly the main factor differentiating the internal structure of the FeAl coatings [8,18,24,25], created by melted or unmelted particles depending on the heat transfer efficiency resulting mainly from the dynamics of the DGS process. As reported by Senderowski et al. [8,18,24,25], the evident composite-like morphology of the FeAl type coating is related to the presence of dispersed intermetallic phases in the various stages of ordering (depending on the Al content). This effect is due to the considerable oxidation in the DGS process, which is the reason for the occurrence of the stable layers of oxides such as α-Al_2O_3 identified as thin films between the nanocrystalline grains of a dual-phase FeAl and Fe_3Al structure.

4. Verification of Model Calculations for Gas Detonation Parameters by DGS Spraying of FeAl Powder Particles into Water

The experimental verification of the model of heat exchange between FeAl particles and the gas flowing around them was based on spraying such particles into water. The model validation included size and chemical composition of powder particles and their susceptibility to melting and deformation under specific DGS process conditions (presented in Section 2).

The performed IPS UA analysis and SEM/BSE observations of FeAl (VIGA) powder in the as-received state confirmed that it is characterized by spherical particles of various sizes in the range of approx. 5–180 µm (Figures 9 and 10). The particle size distribution for each size class identified in the IPS UA study is presented in Figure 9b.

The largest volume of approx. 30.2% of particles with a size of 80 to 125 µm was identified after the test. At the same time, approx. 1.6% by volume share of particles larger than declared by the manufacturer was also found in powder mixture (160–180 µm).

It should be mentioned that only approx. 7.5% of FeAl powder particles (from approx. 320,000 tested particles) possess a diameter in the range from 125 to 160 μm and as much as approx. 60.7% diameter in the range from 5 μm to 80 μm. Finally, the share of the smallest of particles from 5 to 45 μm was approx. 16.6% by volume.

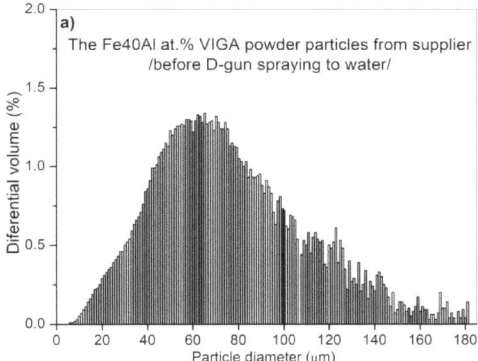
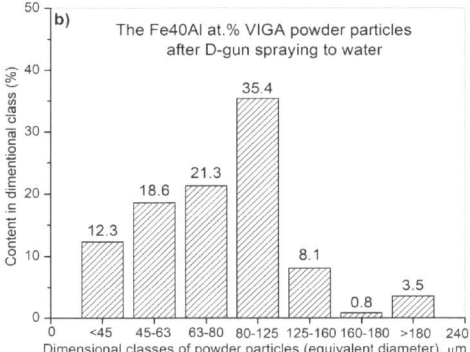

Figure 9. Analysis of the particle diameter distribution of the Fe40Al at.% (VIGA) powder using an IPS UA analyzer in as-received state: volumetric fraction (**a**), fraction of particles in specific size classes (**b**).

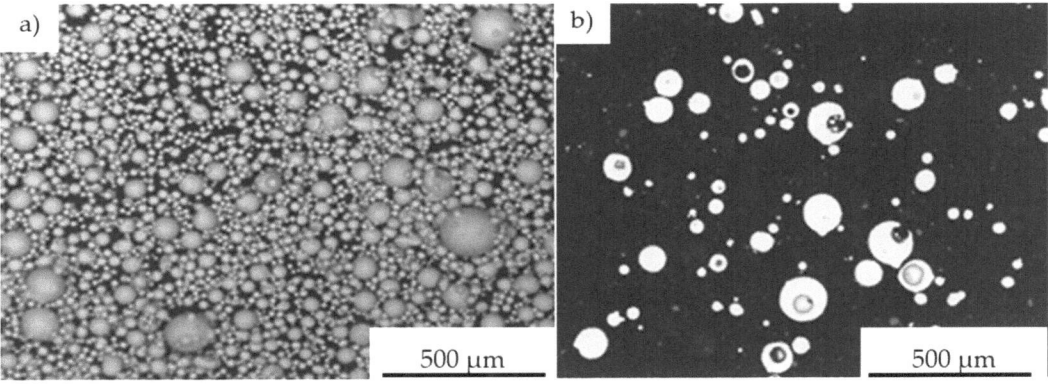

Figure 10. Morphology of Fe40Al at.% powder particles used in D-gun spraying into water: the dimension heterogeneity of the spherically structured particles (**a**), the cross-section reflecting the porosity of FeAl particles in their volume (**b**).

Microscopic observations on the powder particles cross-sections confirmed their high dimensional heterogeneity. These particles were characterized by a spherical shape and high porosity (Figure 11). Such porosity results from the manufacturing process that includes the spraying of the FeAl alloy in an inert gas (argon) previously melted in a VIGA process (vacuum induction melting and inert gas atomization). Detailed investigations on particle porosity showed that the most numerous population (about 22% of 324 identified pores in 250 tested powder particles) were identified as pores with a size in the range of 20–30 μm [8].

The chemical composition microanalysis performed on the cross-sections of FeAl VIGA powder particles in an as-received state confirmed their homogeneity (Figure 11). Their structure was found as a secondary solution containing approx. 40 at.% aluminum, corresponding to the FeAl phase (Table 6).

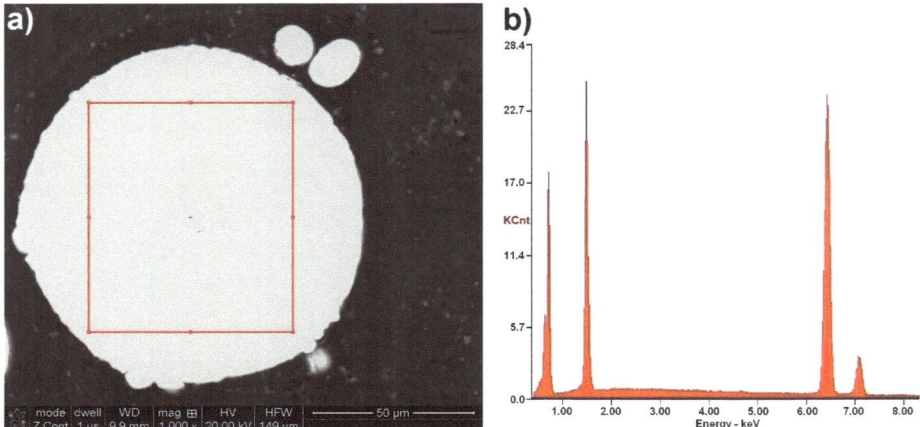

Figure 11. Typical FeAl solid solution microstructure on the cross-section of the FeAl VIGA powder particle (**a**), and SEM/EDX spectrum corresponding to the EDX area analysis (**b**).

Table 6. Chemical composition of FeAl VIGA powder particles.

Element	Content of the Elements	
	wt.%	at.%
Fe	24.72	40.46
Al	75.28	59.54

IPS UA dimensional analysis of FeAl particles sprayed under DGS conditions into water (Figure 12) showed some very large particles with diameter dimension over 180 μm (approx. 3.6%—Figure 13), which did not previously exist in the feedstock, were produced after D-gun spraying. These particles were not found in the as-received powder material (Figure 10). It was found that spraying FeAl particles into water under specific conditions of the DGS process led to an approx. 5.2% increase in the proportion of particles in the 80–125 μm range, with a simultaneous approx. 4.3% decrease in fine particles below 45 μm (Figure 13). A similar decrease of approx. 4.7% was also found in the case of particles in the 45–63 μm range.

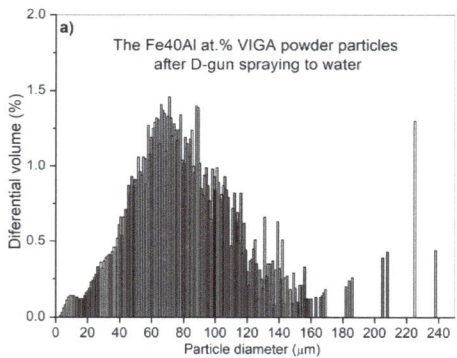
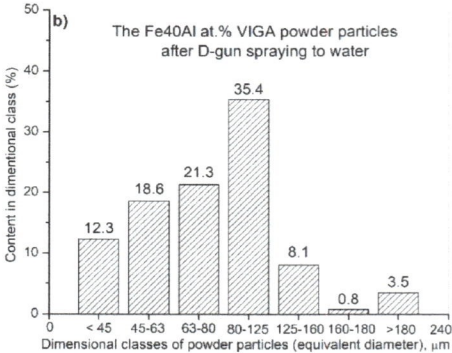

Figure 12. Analysis of the particle diameter distribution of the Fe40Al at.% (VIGA) powder using an IPS UA analyzer after state: volumetric fraction (**a**), fraction of particles in specific size classes (**b**).

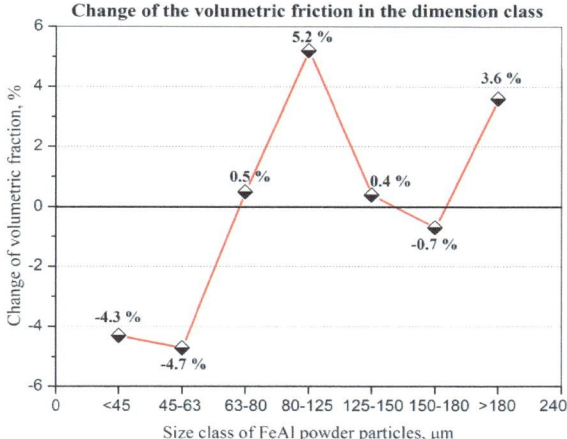

Figure 13. Percentage change in volume fraction in a specific size class of FeAl powder particles sprayed with D-gun into water as compared to as-received FeAl powder.

The SEM/EDS observations reveal, that the change in particle size distribution could be related to the melting of powder particles below 60 μm, which form larger grain clusters upon impact with the water surface (Figure 14). Created agglomerates were formed by the welding of molten fine particles dispersed in a gas stream (Figure 14a). As a result of phase transformations of the FeAl phase melting in the stream of hot gases in the DGS process, larger-size FeAl powders, melted on the surface, can also be welded (Figure 14b,c). This may explain the formation of about 3.6% of FeAl powder particles of diameter above 180 μm, which were not observed in the powder charge introduced into the "Perun-S" detonation gun feeder.

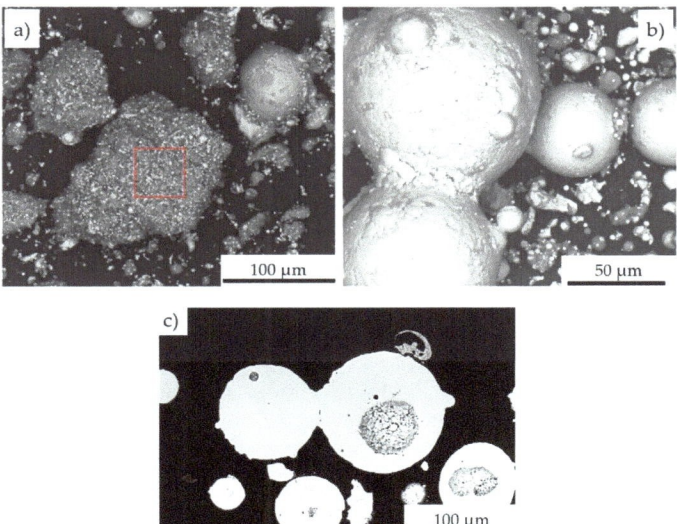

Figure 14. Morphology of Fe40Al at.% powder particles after DGS spraying into water: agglomerate formed from melted particles with the EDS analysis area marked (**a**); welding of two larger FeAl powder particles with larger dimensions (**b**); the cross-section of the FeAl particles welded together with the pores in the particle volume (**c**).

The chemical composition analysis from the surface of five randomly selected grain clusters, formed from melted particles of DGS powder sprayed into water (Figure 14a), presents a significant increase of aluminium and oxygen contents, which due to the lack of SEM/EDS calibration should be treated as a qualitative—semi-quantitative analysis (Table 7).

Table 7. Average results of SEM/EDS chemical composition tests on the surface of agglomerates formed from melted FeAl powder particles under specific DGS spraying conditions.

	Content of the Elements	
	wt.%	at.%
Fe	21.7	9.3 ± 7.2
Al	49.3	45.7 ± 4.1
O	~29	~45 ± 3

The high oxygen content with a significant increase in aluminum content on the surface of agglomerates formed from melted FeAl powder particles confirmed the formation of oxide phases under the DGS conditions. The privileged zones of in situ formation of thin Al_2O_3 films in the structure of gas detonation sprayed FeAl coatings (which are also partially morphized) are presented in other works by one of the authors [8,24,25].

5. Conclusions

In this work, the effect of thermo-kinetic conditions of the gaseous detonation stream on the heating degree and susceptibility to melting of FeAl powder particles during their spraying into water under specific conditions of the DGS process was determined analytically and numerically.

Experimental and modelling analysis on the mechanisms of heat transfer by FeAl powder particles in the stream of gaseous products of detonation combustion, for the DGS spraying process of powder particles into water, enables the following conclusions to be drawn:

1. Gas parameters and thermodynamic state of combustion products at the time of a detonation explosion of the gas mixture in the C-J plane is sufficient to melt FeAl powder particles up to 5 µm in diameter;
2. Under conditions of thermal forcing in the C-J plane, when shock wave passes through the powder particles, the changes in particle velocity are much faster than the particle temperature changes caused by convective heating. Therefore, the main factor affecting the heating and melting degree of FeAl powder particles are gaseous combustion products, detonation during their adiabatic expansion after shock wave passing;
3. Thermal diffusivity of Fe40Al at.% powder particles with a diameter of up to 60 µm (mainly determined by the thermophysical properties of the FeAl phase) does not constitute any barrier to reaching the temperature in the particle volume in accordance with temperature changes on its surface;
4. Powder particles with a diameter of 40 µm will not be able to absorb enough heat to melt them for the determined thermo-gas-kinetic parameters of the gas-detonation stream and the thermo-physical properties of the Fe40Al at.% intermetallic powder charge. Such considerations include the heat conduction in the volume of the FeAl particle and the radiation effects of heat exchange between the FeAl particle and the walls of the detonation gun barrel. On the other hand, for numerical calculations, this diameter is equal to 80 µm;
5. SEM microstructural characterization and particle size distribution of FeAl particles after the DGS spraying process into water revealed an approx. 5.2% increase in the proportion of particles in the 80–125 µm range, and a simultaneous approx. 4.3% decrease in fine particles below 45 µm. A similar decrease of approx. 4.7% was also found in the case of particles in the 45–63 µm range;

6. The analytical model predicts that approximately 16.6% of all particles melt, while the numerical model predicts that nearly 61% of all particles are subjected to melting.

Author Contributions: Conceptualization, C.S. and A.J.P.; methodology, C.S., D.Z., K.V.K., A.J.P. and B.F.; formal analysis, C.S., A.J.P. and M.K.; investigation, C.S., A.J.P., B.F., D.Z. and K.V.K.; writing—original draft preparation, C.S., A.J.P. and M.K.; writing—review and editing, M.K. and C.S.; supervision, C.S.; funding acquisition, C.S. All authors have read and agreed to the published version of the manuscript.

Funding: Financial support from the Polish National Science Centre, Poland, Research Project No. 2015/19/B/ST8/02000, is gratefully acknowledged.

Institutional Review Board Statement: Not applicable.

Informed Consent Statement: Not applicable.

Data Availability Statement: The data are available in a publicly accessible repository.

Acknowledgments: The authors thank Y. Borisov of the E.O. Paton Electric Welding Institute of the National Academy of Sciences of Ukraine in Kiev for enabling the DGS experiments.

Conflicts of Interest: The authors declare no conflict of interest.

References

1. Ulianitsky, V.Y.; Dudina, D.V.; Shtertser, A.A.; Smurov, I. Computer-controlled detonation spraying: Flexible control of the coating chemistry and microstructure. *Metals* **2019**, *9*, 1244. [CrossRef]
2. Ulianitsky, V.; Shtertser, A.; Zlobin, S.; Smurov, I. Computer-controlled detonation spraying: From process fundamentals toward advanced applications. *J. Therm. Spray Technol.* **2011**, *20*, 791–801. [CrossRef]
3. Dudina, D.V.; Batraev, I.S.; Ulianitsky, V.Y.; Korchagin, M.A. Possibilities of the computer-controlled detonation spraying method: A chemistry view point. *Ceram. Int.* **2014**, *40*, 3253–3260. [CrossRef]
4. Rybin, D.K.; Batraev, I.S.; Dudina, D.V.; Ukhina, A.V.; Ulianitsky, V.Y. Deposition of tungsten coatings by detonation spraying. *Surf. Coat. Technol.* **2021**, *409*, 126943. [CrossRef]
5. Nikolaev, Y.A.; Vasil'ev, A.A.; Ul'anitskii, B.Y. Gas detonation and its application in engineering and technologies (Review). *Combust. Explos. Shock Waves* **2003**, *39*, 382–410. [CrossRef]
6. Batraev, I.S.; Ulianitsky, V.Y.; Dudina, D.V. Detonation spraying of copper: Theoretical analysis and experimental studies. *Mater. Today Proc.* **2017**, *4*, 11346–11350. [CrossRef]
7. Senderowski, C.; Bojar, Z. Influence of Detonation Gun Spraying Conditions on the Quality of Fe-Al Intermetalic Protective Coatings in the Presence of NiAl and NiCr Interlayers. *J. Therm. Spray Technol.* **2009**, *18*, 435–447. [CrossRef]
8. Senderowski, C. *Żelazowo-Aluminiowe Intermetaliczne Systemy Powłokowe Uzyskiwane z Naddźwiękowego Strumienia Metalizacyjnego (Iron-Aluminium Intermetallic Coatings Synthesized by Supersonic Metallization Stream)*; Bel Studio: Warsaw, Poland, 2015; 280p, ISBN 978-83-7798-227-3. (In Polish)
9. Kadyrov, E. Gas-particle interaction in detonation spraying systems. *J. Therm. Spray Technol.* **1996**, *5*, 185–195. [CrossRef]
10. Kadyrov, E.; Kadyrov, V. Gas dynamical parameters of detonation powder spraying. *J. Therm. Spray Technol.* **1995**, *4*, 280–286. [CrossRef]
11. Ramadan, K.; Butler, P.B. Analysis of gas flow evolution and shock wave decay in detonation thermal spraying systems. *J. Therm. Spray Technol.* **2004**, *13*, 239–247. [CrossRef]
12. Ramadan, K.; Butler, P.B. Analysis of particle dynamics and heat transfer in detonation thermal spraying systems. *J. Therm. Spray Technol.* **2004**, *13*, 248–264. [CrossRef]
13. Korytchenko, K.V.; Tomashevskiy, R.S.; Varshamova, I.S.; Meshkov, D.V.; Samoilenko, D. Numerical investigation of energy deposition in spark discharge in adiabatically and isothermally compressed nitrogen. *Jpn. J. Appl. Phys.* **2020**, *59*, 1–11. [CrossRef]
14. Fikus, B.; Senderowski, C.; Panas, A.J. Modeling of Dynamics and Thermal History of Fe40Al Intermetallic Powder Particles Under Gas Detonation Spraying Using Propane–Air Mixture. *J. Therm. Spray Technol.* **2019**, *28*, 346–358. [CrossRef]
15. Senderowski, C.; Bojar, Z.; Łodziński, K.; Morgiel, J. Principal mechanisms of the structure formation in Fe-Al type d-gun intermetallic coatings. In Proceedings of the 20th Congress of International Federation for Heat Treatment and Surface Engineering, Beijing, China, 23–25 October 2012; pp. 746–751.
16. Astakhov, E.A. Controlling the properties of detonation-sprayed coatings: Major aspects. *Powder Metall. Met. Ceram.* **2008**, *47*, 70–79. [CrossRef]
17. Palm, M. Concepts derived from phase diagram studies for the strengthening of Fe–Al-based alloys. *Intermetallics* **2005**, *13*, 1286–1295. [CrossRef]
18. Panas, A.J.; Senderowski, C.; Fikus, B. Thermophysical properties of multiphase Fe-Al intermetallic-oxide ceramic coatings deposited by gas detonation spraying. *Thermochim. Acta* **2019**, *676*, 164–171. [CrossRef]

19. Wolczynski, W.; Senderowski, C.; Morgiel, J.; Garzel, G. D-gun sprayed Fe-Al single particle solidification. *Arch. Metall. Mater.* **2014**, *59*, 211–220. [CrossRef]
20. Pawlowski, A.; Senderowski, C.; Wolczynski, W.; Morgiel, J.; Major, L. Detonation deposited Fe-Al coatings part II: Transmission electron microscopy of interlayers and Fe-Al intermetallic coating detonation sprayed onto the 045 steel substrate. *Arch. Metall. Mater.* **2011**, *56*, 71–79. [CrossRef]
21. Pawlowski, A.; Czeppe, T.; Major, L.; Senderowski, C. Structure morphology of Fe-Al coating detonation sprayed onto carbon steel substrate. *Arch. Metall. Mater.* **2009**, *54*, 783–788.
22. Senderowski, C.; Pawlowski, A.; Bojar, Z.; Wolczynski, W.; Faryna, M.; Morgiel, J.; Major, L. TEM microstructure of Fe-Al coatings detonation sprayed onto steel substrate. *Arch. Metall. Mater.* **2010**, *55*, 373–381.
23. Senderowski, C.; Bojar, Z. Gas detonation spray forming of Fe-Al coatings in the presence of interlayer. *Surf. Coat. Technol.* **2008**, *202*, 3538–3548. [CrossRef]
24. Senderowski, C.; Bojar, Z.; Wołczyński, W.; Pawłowski, A. Microstructure characterization of D-gun sprayed Fe–Al intermetallic coatings. *Intermetallics* **2010**, *18*, 1405–1409. [CrossRef]
25. Senderowski, C. Nanocomposite Fe-Al intermetallic coating obtained by Gas Detonation Spraying of mlled self-decomposing powder. *J. Therm. Spray Technol.* **2014**, *23*, 1124–1134. [CrossRef]
26. Fauchais, P.L.; Heberlain, V.R.; Boulos, M.I. *Thermal Spray Fundamentals. From Powder to Part*; Springer: New York, NY, USA, 2014.
27. Sova, A.; Pervushin, D.; Smurov, I. Development of multimaterial coatings by Cold spray and gas detonation spraying. *Surf. Coat. Technol.* **2010**, *205*, 1108–1114. [CrossRef]
28. Ulianitsky, V.Y.; Dudina, D.V.; Batraev, I.S.; Kovalenko, A.I.; Bulina, N.V.; Bokhonov, B.B. Detonation spraying of titanium and formation of coatings with spraying atmosphere-dependent phase composition. *Surf. Coat. Technol.* **2015**, *261*, 174–180. [CrossRef]
29. Senderowski, C.; Panas, A.J.; Paszula, J.; Bojar, Z. Ocena stopnia ogrzania cząstek proszku FeAl w procesie natryskiwania gazodetonacyjnego (GDS) (The Evaluation of FeAl Particle Thermal Response in Gas Detonation Thermal Spraying Process). *Mater. Eng.* **2013**, *6*, 849–853. (In Polish)
30. Wiśniewski, S. *Wymiana Ciepła*; PWN: Warszawa, Poland, 1994. (In Polish)
31. Kharlamov, Y.A. Detonation spraying of protective coatings. *Mater. Sci. Eng.* **1987**, *93*, 1–37. [CrossRef]
32. Tarzhanov, V.I.; Telichko, I.V.; Vil'danov, V.G.; Sdobnov, V.I.; Makarov, A.E.; Mukhin, S.L.; Koretskii, I.G.; Ogarkov, V.A.; Vlasov, V.V.; Zinchenko, A.D.; et al. Detonation of Propane-Air Mixtures under injection of hot detonation products. *Combust. Explos. Shock Waves.* **2006**, *3*, 336–345. [CrossRef]
33. Gavrilenko, T.P.; Nikolaev, Y.A. Calculation of detonation gas spraying. *Combust. Explos. Shock Waves.* **2007**, *43*, 724–731. [CrossRef]
34. Li, M.; Yan, C.; Zheng, L.; Wang, Z.; Qiu, H. Investigations on multicycle spray detonations. *Front. Energy Power Eng. China* **2007**, *1*, 207–212. [CrossRef]
35. Glassman, I.; Yetter, R.A. *Combastion*, 4th ed.; Elsevier Inc.: Amsterdam, The Netherlands, 2008.

Article

Phase Structure Evolution of the Fe-Al Arc-Sprayed Coating Stimulated by Annealing

Tomasz Chmielewski [1,*], Marcin Chmielewski [2], Anna Piątkowska [2], Agnieszka Grabias [2], Beata Skowrońska [1] and Piotr Siwek [1]

[1] Institute of Manufacturing Technologies, Warsaw University of Technology, Narbutta Str. 85, 02-524 Warsaw, Poland; beata.skowronska@pw.edu.pl (B.S.); siwek_piotr@wp.pl (P.S.)

[2] Łukasiewicz Research Network—Institute of Microelectronics and Photonics, Al. Lotników 32/46, 02-668 Warsaw, Poland; Marcin.Chmielewski@imif.lukasiewicz.gov.pl (M.C.); Anna.Piatkowska@imif.lukasiewicz.gov.pl (A.P.); Agnieszka.Grabias@imif.lukasiewicz.gov.pl (A.G.)

* Correspondence: tomasz.chmielewski@pw.edu.pl; Tel.: +48-22-849-9797

Citation: Chmielewski, T.; Chmielewski, M.; Piątkowska, A.; Grabias, A.; Skowrońska, B.; Siwek, P. Phase Structure Evolution of the Fe-Al Arc-Sprayed Coating Stimulated by Annealing. *Materials* **2021**, *14*, 3210. https://doi.org/10.3390/ma14123210

Academic Editor: Cezary Senderowski

Received: 27 February 2021
Accepted: 3 June 2021
Published: 10 June 2021

Publisher's Note: MDPI stays neutral with regard to jurisdictional claims in published maps and institutional affiliations.

Copyright: © 2021 by the authors. Licensee MDPI, Basel, Switzerland. This article is an open access article distributed under the terms and conditions of the Creative Commons Attribution (CC BY) license (https://creativecommons.org/licenses/by/4.0/).

Abstract: The article presents the results of research on the structural evolution of the composite Fe-Al-based coating deposited by arc spray with initial low participation of in situ intermetallic phases. The arc spraying process was carried out by simultaneously melting two different electrode wires, aluminum and low alloy steel (98.6 wt.% of Fe). The aim of the research was to reach protective coatings with a composite structure consisting of a significant participation of Fe_xAl_y as intermetallic phases reinforcement. Initially, synthesis of intermetallic phases took place in situ during the spraying process. In the next step, participation of Fe_xAl_y fraction was increased through the annealing process, with three temperature values, 700 °C, 800 °C, and 900 °C. Phase structure evolution of the Fe-Al arc-sprayed coating, stimulated by annealing, has been described by means of SEM images taken with a QBSD backscattered electron detector and by XRD and conversion electron Mössbauer spectroscopy (CEMS) investigations. Microhardness distribution of the investigated annealed coatings has been presented.

Keywords: Fe-Al type intermetallics; phase synthesis; arc sprayed coatings

1. Introduction

Transition iron aluminides are attractive coating materials with specific properties, especially in comparison with nickel or chromium-based materials. Fe-Al intermetallic systems have been among the most intensively studied over the last few decades [1]. Many intermetallic applications of the Fe-Al system relate to protective coatings made by different methods, like laser cladding [2,3], D-gun spray [4–6], flame spray [7,8], arc spray [9–11], cold gas spraying [12], and plasma transferred arc cladding [13], some using the strategy of employing the elemental powder materials [14]. Intermetallic phases, due to their advantages, are increasingly often used as a surface material, whose purpose is to work at high temperatures [15,16]. They are significantly resistant to oxidation, carburizing, and sulfation at high temperatures (up to 900 °C) [17,18]. Additionally, they are highly resistant to erosion [19] and cavitation [20,21], and have relatively low density and low prices compared to corrosion-resistant [22,23] and acid-resistant steel [24], which require application expensive elements, such as Cr, Ni, Mo [25–27]. The intermetallics owe their special properties to their ordered structures with strong chemical bonds and simultaneous dense packing of atoms in crystal lattices, which leads to reduced diffusion velocity, creep resistance, and resistance to high-temperature corrosion [28–31].

There are reports in the literature that have described the processes of in situ manufacturing of intermetallic phases on a surface layer by the alloying of components [32,33]. Numerous cases of annealing-stimulated intermetallic synthesis are known in the literature [34–37], which could also be used as a part of the joining procedure [22]. The main disadvantage that limits the use of Fe-Al intermetallics is their brittleness at room temperature and the difficulty of shaping ready-made elements to desired dimensions by means of mechanical machining methods [38,39]. Due to the significant differences in properties like melting point and specific heat of Fe and Al, it is difficult to obtain materials with a reproducible composition and homogeneous structure [40].

The method of producing a composite material using the Fe-Al intermetallic phase proposed in the article is a continuation of the procedure proposed in [10] and may constitute an alternative to the currently used solutions, which are usually much more expensive and based mainly on ready-to-use intermetallic powders [41,42].

2. Materials and Experimental Procedure

The initial investigation [10] and its results showed that a 0.5 mm thick and dense Fe-Al composite coating with uniformly distributed Fe and Al particles can be deposited by arc spraying onto the substrate of an S355 JR steel plate ($50 \times 100 \times 5$ mm^3). The coating obtained included two main metallic phases based on the Fe (bcc) and Al (fcc) structure. However, Fe-Al precipitations have been revealed on the boundary grains as well as on the Al and Fe matrix. Volume fraction of intermetallic phases in the sprayed coating was too low to use XRD method to investigate them. Results showed that both SEM and EDS analyses confronted with Mössbauer spectroscopy analysis confirmed presence of FeAl intermetallic phases in the structure with varying atomic factors of iron and aluminum, including approximately 50-50% and 80-20% at the structure of the coating. Volume fraction of Fe$_x$Al$_y$ phases was between 7% and 10%.

In this paper, an iron/aluminum composite coating deposited by arc spraying using iron and aluminum wires in the condition described in [10] was annealed at different temperature (700 °C, 800 °C, and 900 °C) values for 2 h to aim at increasing the volume fraction of iron aluminide intermetallic in composite coating. The parameters of the thermal annealing cycle were selected on the basis of a literature analysis [43–45]. Annealing processes have been realized in vacuumed atmospheres (5×10^{-5} Tr) with an experimental vacuum chamber with induction heating, designed and built in the welding department of Warsaw University of Technology (Poland). After annealing, the samples were cooled and prepared for metallographic examination in the cross-section and on the surface. The specimens were grounded in the cross-section and on the coating's surface with abrasive paper up to 2000 grits, polished to obtain a mirror-finished surface for microstructure observation and chemical and phase composition analysis. The composite, multi-phase Fe-Al arc-sprayed coating on the steel substrate after annealing was analyzed via scanning electron microscope Auriga produced by the Zeiss company (Oberkochen, Germany). The microstructure and phase structure of the coating after annealing were investigated by using SEM, XRD, and Mössbauer spectroscopy analogously to the study its structure before annealing for comparison. X-ray diffraction measurements were performed using the Siemens D500 powder diffractometer (Siemens, Munich, Germany), equipped with a high-resolution semiconductor Si:Li detector and Kα1,2Cu radiation λ = 1.5418 Å. Conversion electron Mössbauer spectroscopy (CEMS) has been carried out using a constant acceleration home-made Mössbauer spectrometer (designed and built by Łukasiewicz Research Network—Institute of Microelectronics and Photonics) with integrated gas flow electron counter. Measurements were performed at room temperature with the use of a ^{57}Co-in-Rh source. The annealed coatings were probed up to about 200 nm in depth. As a supplement to the assessment of structural changes, the distribution of the coating's hardness in the cross-section along four parallel lines up to the surface layer of substrate material were performed according to the Vickers method using the Leitz–Wetzlar microhardness tester (LEICA, Wetzlar, Germany) (load 100 g for 10 s).

3. Results and Discussion

3.1. SEM Investigation of Annealed Coatings

SEM images taken throughout the entire coating (Figure 1) revealed a large variety of phases and their generally lamellar distribution nature, specific to thermally sprayed coatings [46,47]. The aluminum wire is chemically active in the arc spraying process, such that melted particles are already oxidized in metallization stream and the obtained coatings always contain oxide films, mainly present in the areas between of lamellar grains of higher aluminum content.

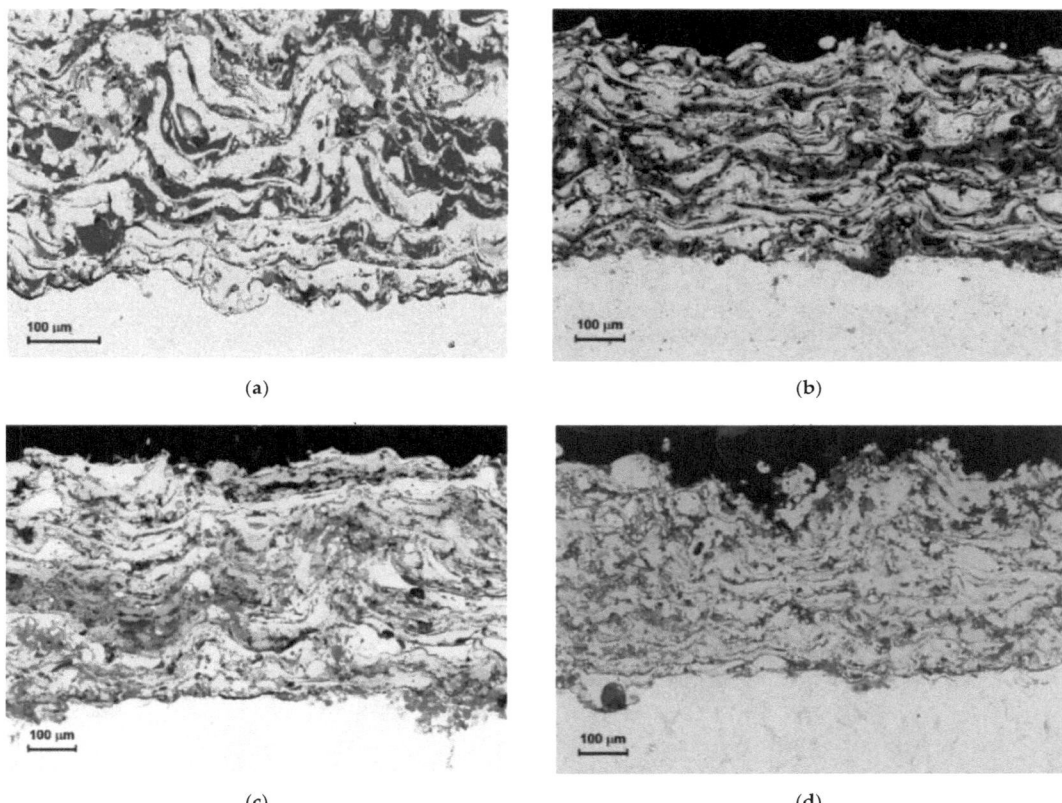

Figure 1. SEM/QBSE images of as-arc-sprayed Fe-Al-type coating (**a**), and after annealing for 2 h, at respectively: (**b**) 700 °C, sample 1; (**c**) 800 °C, sample 2; (**d**) 900 °C, sample 3.

As a result of high kinetic energy, temperature, and the speed of melted particles in the arc-spraying process, the Fe-Al-type coatings that were about 500 μm thick showed a lamellar microstructure with an inhomogeneous phases distribution where a few spherical non-molten particles were visible (Figure 1a). Based on SEM/QBSE images, different shades of gray were observed for indicated phases with different atomic masses in the backscattered electron (QBSE) detector.

Various shades of gray areas in the SEM/QBSE images correspond to different phases—predominant Fe(Al) solid solution with high iron is indicated by the lightest grains and the aluminum-rich phases, including oxides and spinels, are represented by the darkest grains in Figure 1a.

The local occurrences of disordered and dispersed intermetallic Fe-Al-type phases in various Al content must be also emphasized (presented as the varied degree of grayness in individual grains in the QBSE examination and slightly less bright than these high-Al grains—Figure 1a).

The high degree of chemical inconsistency of the arc-sprayed Fe-Al-type coating also proves its intermetallic phase inhomogeneity, especially when one takes into consideration the fact that a big range of acceptable changes in the alloying elements contents in the Fe-Al intermetallic phase solutions [46,47]. The comparative analysis in the cross-sections of the Fe-Al type coating (Figure 1) indicate that the Fe-Al arc-sprayed coating (Figure 1a) is not uniform in the phase structure and morphology throughout its thickness [10]. However phase structure of the coating has been significantly changed, in the direction of higher homogeneity, after annealing sample 1 at 700 °C (Figure 1b), sample 2 at 800 °C (Figure 1c), and sample 3 at 900 °C (Figure 1d).

As the Fe-Al type arc-sprayed coatings were annealing at high temperatures, the thermal and hardness stability of the Fe-Al type intermetallic coating (such as the change in the morphology and chemical composition of the lamellar structure, phase change susceptibility, and the degree of strengthening of the coating) was analyzed after heating at a high-temperature (respectively 700 °C, 800 °C, and 900 °C) for 2 h.

3.2. EDX Measurements

3.2.1. Arc-Sprayed Fe-Al-Type Coating Annealed at 700 °C

The basis for identification of the structural analysis and inhomogeneity of the chemical composition (phase composition) in the arc-sprayed Fe-Al type coatings after they were annealed at high temperatures were the EDX results of point and linear microanalysis of the chemical composition, as well as mapping of Fe, Al, and O elements.

Based on the results of SEM/EDX point microanalysis, it was shown (Figure 2) that the layered arrangement of grains of the arc Fe-Al-type coating annealed at 700 °C revealed a considerably varied chemical composition with an extended range of solid solution from about 6 to 37 at.% Al (Table 1). This wide range of compositions implies the occurrence of grains based on the low-aluminum Fe(Al) solid solution grains observed as the bright gray areas in the BSE image (p1 and p2 in Figure 2).

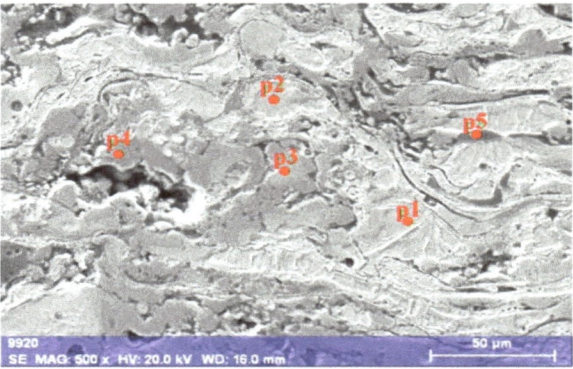

Figure 2. Microstructure of arc-sprayed Fe-Al-type intermetallic coating after annealing at 700 °C for 2 h with hypothetical phase identification based on results of EDS point analysis (in Table 1).

Table 1. Chemical composition of as-arc-sprayed Fe-Al-type coating after annealing at 700 °C/2 h based EDX point microanalysis according to Figure 2.

Designation of Grain Area According to Figure 2	Content, at. %							
	C	O	Al	Si	Mn	Fe	Au	Occurrence
p1	~9	~1	6.47	0.48	0.66	81.49	0.15	often
p2	~10	-	7.69	0.84	0.60	80.62	0.13	often like p1
p3	~12	~19	24.90	0.55	0.72	41.85	0.10	rare
p4	~4	~51	6.33	0.29	0.38	36.25	0.09	rare like p3
p5	~15	~35	35.76	0.14	0.64	12.87	0.09	very rare

Annealing at 700 °C for 2 h also caused a local presence of disordering secondary solution based on Fe-Al phases ranging from 25 to 36 at.% Al and a significant content of oxygen (p3 and p5 in Figure 2). However, near some Fe-Al phases, depletion of Al entailed a local occurrence of oxides in the form of spinels (p4 in Figure 2), as well as aluminum oxides created between lamellar grains (p5 in Figure 2) where Fe was doped from the matrix of Fe-Al intermetallic material.

In the discussion of the EDX analysis, the identified trace elements were not taken into account (Table 1), as the contaminations of Si, Mn, and Au could be caused by polishing of the samples as well as sputtering with a thin Au layer about 10 nm thick in order to avoid issues with the electrical charge of the primary electron beam in the preparation of the cross-sections of the samples that were obtained using a non-conductive resin as a matrix.

The relative content of these contaminations was below 1 at.% and can be considered as negligible. In the case of light elements such as oxygen and carbon, the analysis was semi-quantitative. The significant presence of carbon was most probably related to the preparation of cross-sections of the samples, where during the polishing process the resin could be transferred to the studied surface of the sample.

The basis for identification of oxides in the Fe-Al-type coating after annealing at 700 °C were the results of a linear microanalysis of the chemical composition (Figure 3), as well as mapping of Fe, Al, and O elements (Figure 4), which proved that oxygen, apart from oxide spinels, is mainly present in the areas of grains of higher aluminum content (imaged as dark grey and grey), but also in the light grey areas identified as disordering secondary solid solution based on Fe-Al phases with decreased Al content.

Additionally, the results of the linear EDS measurements (Figure 3a,b) performed on the representative surface of the Fe-Al-type coating annealed at 700 °C showed very different proportions of the Al-Fe elemental ranging from 10 to 50 at.% Al along the line with a length of approx. 20 µm. The linear EDX results of changes in the proportions and chemical compositions at the grain boundary cross-sections of the coatings are shown in Figure 3c,d. In the region marked with the yellow arrow, the phase composition of roughly 50 at.% Fe was observed up to about first 10 µm. The measurement along the green arrow indicates at one end Fe oxide, whereas on the other end there is Al oxide. In the middle, a mixed Fe-Al-O phase is observed.

This is a confirmation of the fact that the lamellar grains created in the arc-spraying conditions after annealing at 700 °C exhibited certain composite features due to the different Fe-Al type phases as components of the structure inherited from the arc-spraying wire material, only with inconsiderable evidence of the phase transformation during heating at 700 °C for 2 h. The aluminum wire is chemically active in the arc-spray process, such that all melted particles are already oxidized in the obtained coatings and always contain oxide films inside the coating and at the internal interfaces. It is mainly formation of the oxide films (the oxidized blue areas of the coating structure on EDX maps, Figure 4a) that brings phase transformation during heating at 700 °C for 2 h.

3.2.2. Arc-Sprayed Fe-Al-Type Coating Annealed at 800 °C

After heat treatment at 800 °C for 2 h, the SEM/EDX results (Figures 5 and 6) revealed the inhomogeneous lamellar structure characteristic-like arc spraying with a varied chemical composition based on the Fe-Al-type phases and oxides, identified mainly in the interlamellar grain boundaries of the coating (Figures 5 and 6). Based on the results of scanning electron microscopy and point EDX microanalysis and mapping, it was shown (Figures 5 and 6a,b) that the grains based on the Fe-Al phases had the range of secondary solid solution from ~14 to ~33 at.% Al (Table 2).

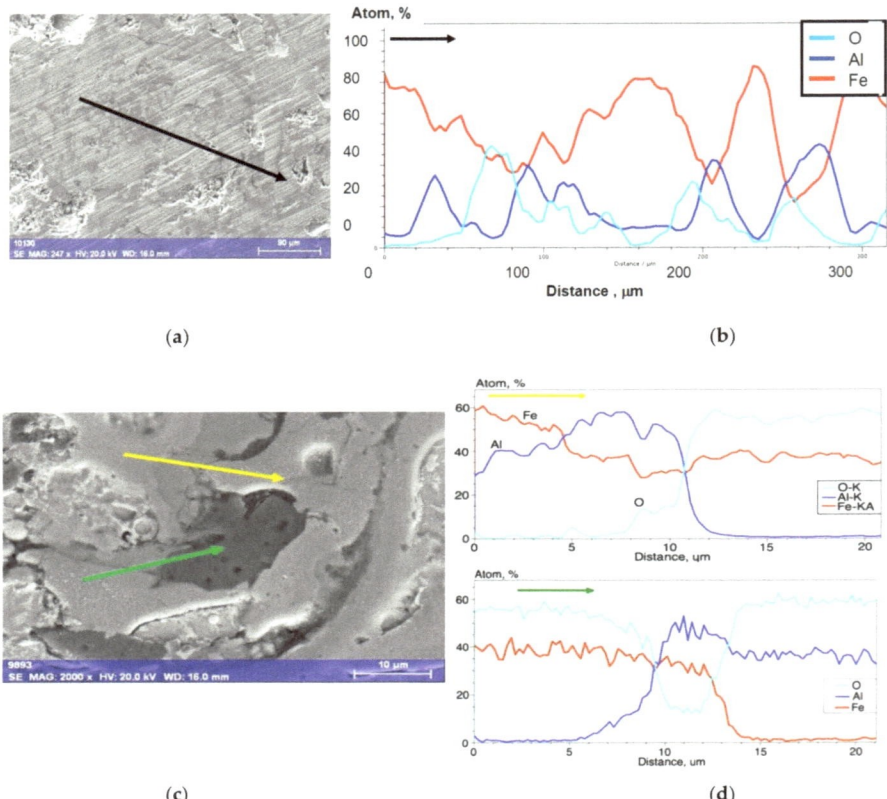

Figure 3. Linear chemical analysis of as-arc-sprayed Fe-Al-type coating after annealing at 700 °C/2 h: (**a**) SE image of the surface of the coating, (**b**) linear distribution of elements along the black arrow marked on Figure 3a, (**c**) SE image with details of the layer cross-section, (**d**) linear distribution of elements along the black narrow marked on Figure 3a, (c) SEM image with details of the layer cross-section, (**d**) linear distribution of elements along the yellow and green arrows marked on Figure 3c.

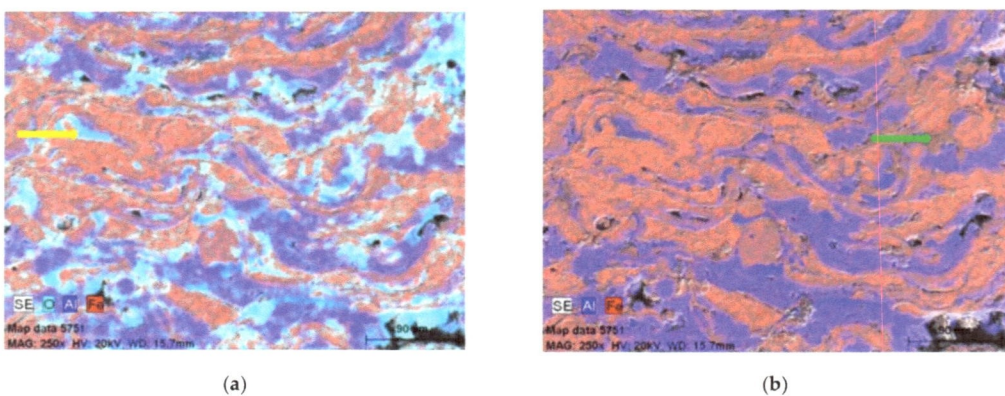

Figure 4. EDX maps of elemental distribution in cross-section of the Fe-Al-type coating annealed at 700 °C: (**a**) O, Al, Fe; (**b**) Al, Fe.

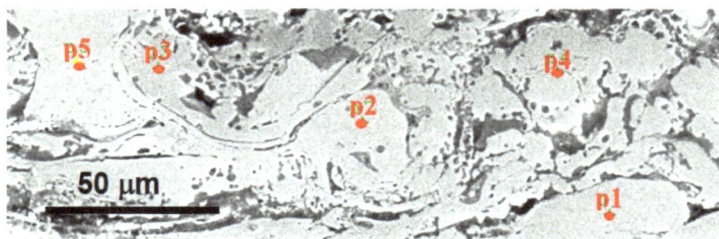

Figure 5. Microstructure of arc-sprayed Fe-Al-type intermetallic coating after annealing at 800 °C for 2 h with hypothetical phase identification based on results of EDX point analysis (in Table 2).

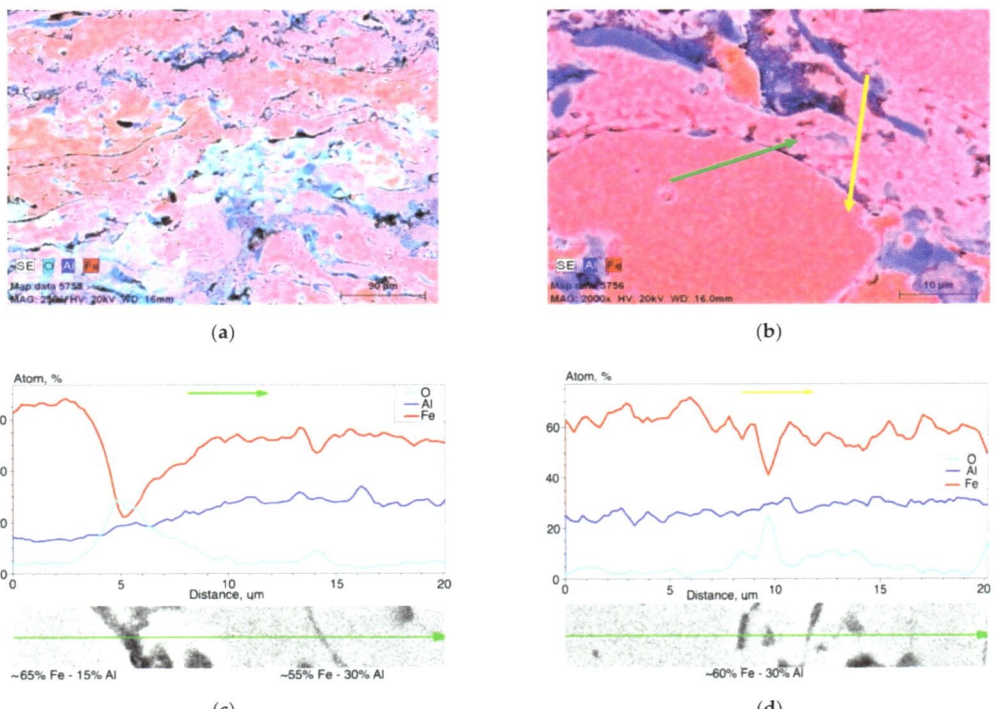

Figure 6. EDX results of elements distribution in cross-section of the coating annealed at 800 °C: maps of (**a**) O, Al, and Fe and (**b**) Al and Fe, as well as plots of the linear distribution of elements along two lines, respectively—(**c**) green and (**d**) yellow—as marked on (**b**).

Table 2. Chemical composition of as-arc-sprayed Fe-Al-type coating after annealing at 800 °C/2 h based on EDX point microanalysis according to Figure 5.

Designation of Grain Area According to Figure 5.	Content, at.%							
	C	O	Al	Si	Mn	Fe	Au	Occurrence
p1	~9	-	21.60	0.92	0.57	67.49	0.08	often
p2	~8	~3	17.34	0.70	0.39	69.55	0.07	often
p3	~9	-	32.86	0.40	0.32	56.63	0.21	rare
p4	~9	-	25.84	0.74	0.53	63.18	0.09	often
p5	~13	~8	13.93	0.63	0.43	62.99	0.26	medium

The share of the dark phase related to Al and the Fe-Al intermediate phases were increased. Areas in the pink color in Figure 6a,b were suitable for Fe-Al phases. The more violet the color of the surface in the structure, the greater the proportion of aluminum. The SEM/EDX results shown in Figures 5 and 6 and Table 2 allowed identification of the fact that in the Fe-Al-type coating annealed at 800 °C/2 h, the most common phase was low-aluminum Fe_3Al at.%, as shown by the disordered secondary solid solution observed in the brightest areas of the SEM/BSE image (Figure 5) and in the pink Fe- and Al-mapping (Figure 6a,b). In the plots of the variations of the aluminum, oxygen, and iron in the linear EDX analyses (Figure 6c,d), the peak oxygen oscillations corresponded to the passage of the analyzing beam of electrons through the oxides, which was the inherent structural component of the Fe-Al-type coating annealed at 800 °C/2 h.

3.2.3. Arc-Sprayed Fe-Al-Type Sample 3 Annealed at 900 °C

The morphology of arc-sprayed Fe-Al-type coating annealed at 900 °C for 2 h was related to the presence of dispersed intermetallic phases in the various stages of ordering (depending on the Al content), as shown in Figures 7 and 8, where the different phases on the cross-section of the coating were revealed.

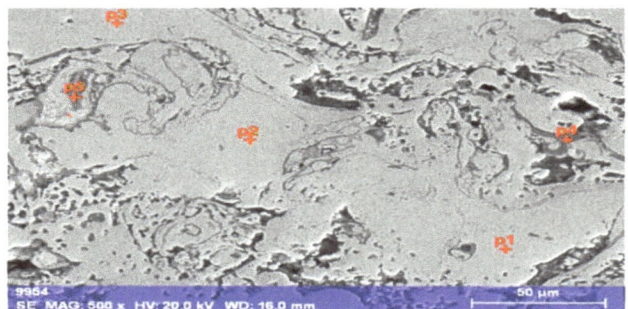

Figure 7. Microstructure of arc-sprayed Fe-Al-type intermetallic coating after annealing at 900 °C for 2 h with hypothetical phase identification based on results of EDS point analysis (in Table 3).

Table 3. Chemical composition of as-arc-sprayed Fe-Al-type coating after annealing at 900 °C/2 h based EDX point microanalysis according to Figure 7.

Designation of Grain Area According to Figure 7	Content, at.%							
	C	O	Al	Si	Mn	Fe	Au	Occurrence
p1	~15	-	21.86	0.55	0.64	61.78	0.10	often
p2	~9	-	21.64	0.79	0.66	67.67	0.15	often like p1
p3	~8	-	23.57	0.48	0.63	67.11	0.14	often like p1
p4	~6	~56	34.62	0.21	0.94	1.22	0.05	rare -precipitation
p5	~4	~55	34.47	0.04	0.23	4.47	0.08	rare

After heat treatment, the microstructure lost its lamellar-like structure, which appeared to have more homogeneity at this point but was still not quite regular. In the middle area of the EDX map in Figure 8a, the phase indicated by the red color was rich in Fe and, it coexisted with the dominant Fe-Al phase in bright weight contrast marked in pink in the EDX Fe and Al maps. On the other hand, the Al-rich phase occurred sporadically, mainly crystallized as narrow lamellas between the matrix areas, and it was also highly oxidized, as presented on Figure 8c.

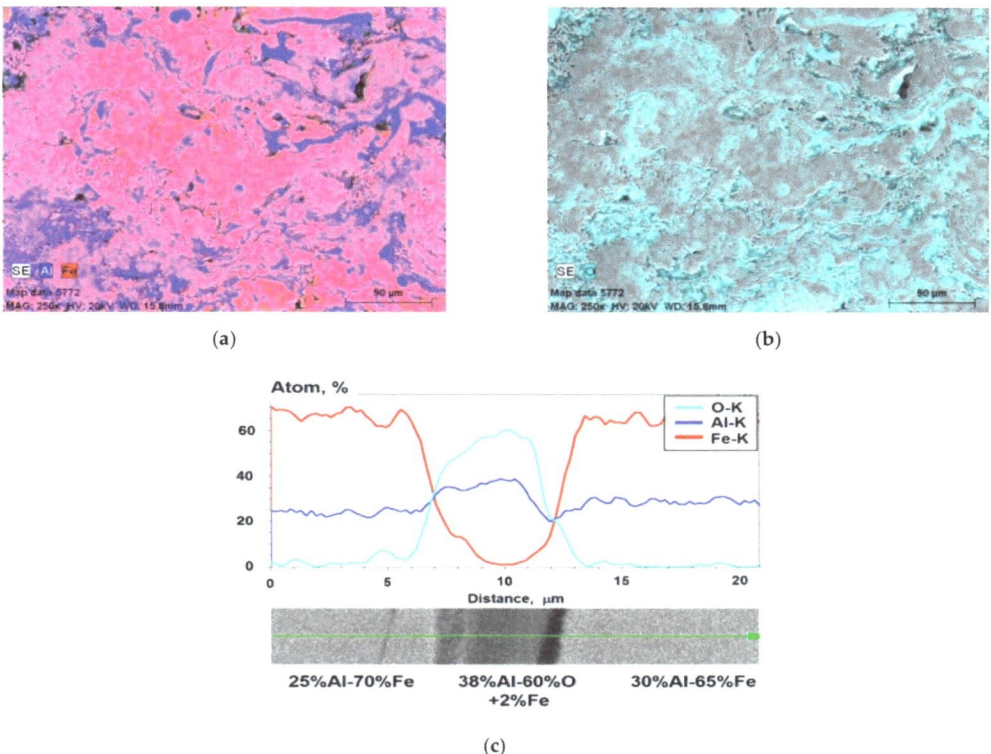

Figure 8. EDX results of elemental distribution in cross-section of the coating annealed in 900 °C: maps of (**a**) Al and Fe and (**b**) O, as well as (**c**) plots of the linear distribution of elements along green line.

3.3. XRD Analysis

In order to analyze the phase structure of the coatings, the XRD analysis was performed for three representative samples annealed at the temperatures of 700 °C, 800 °C, and 900 °C.

Figure 9 presents a diffraction pattern that implies that the phase structure of the studied coating after annealing at 700 °C consisted mainly of the metallic phase based on the bcc Fe phase, which was accompanied by the intermetallic Al_5Fe_2 phase. The dominant bcc phase was most probably an Fe(Al) solid solution with a substantial content of the solvent, which was suggested by a significantly larger value of the lattice constant as compared to pure Fe. A small fraction of an iron oxide phase with the wustite-like structure (FeO) was also seen in the XRD pattern. The volume fraction of this phase in the structure was approximately 5%; however, after annealing at 800 and 900 °C, it dropped below the level of XRD detection.

The XRD pattern obtained for the coating annealed at 800 °C is shown in Figure 10. The phase structure of the analyzed coating consisted of a substantial fraction of the FeAl intermetallic phase, as well as the bcc Fe-based metallic phase and a metastable $Al86Fe14$ phase. Figure 11 presents the diffraction pattern of the sample annealed at 900 °C, which revealed two main phases. The dominant intermetallic FeAl phase coexisted with the bcc Fe-based metallic phase.

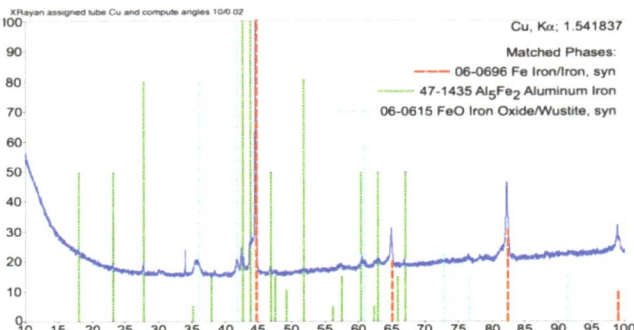

Figure 9. XRD diffraction pattern of the arc-sprayed Fe-Al coating after 700 °C annealing.

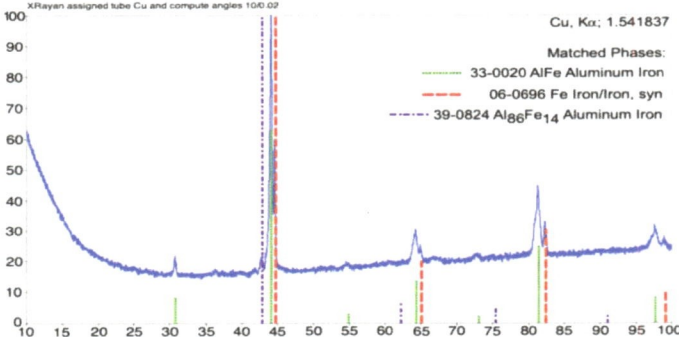

Figure 10. XRD diffraction pattern of the arc-sprayed Fe-Al coating after 800 °C annealing.

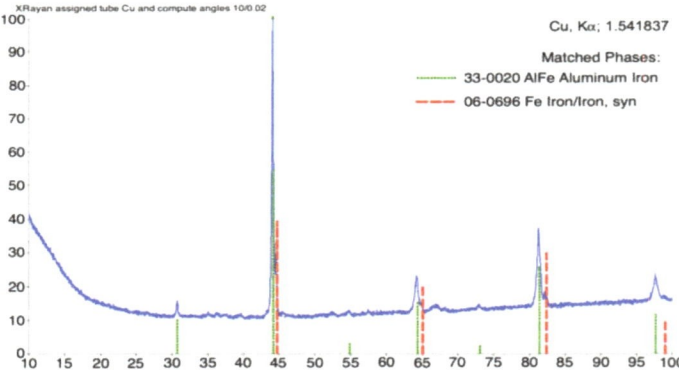

Figure 11. XRD diffraction pattern of the arc-sprayed Fe-Al coating after 900 °C annealing.

The XRD studies were complemented by conversion electron 57Fe Mössbauer spectroscopy measurements, which provided additional information regarding the atomic environment of iron atoms in the studied coatings.

3.4. Mössbauer Spectroscopy Results of Annealed Coatings

The identification of iron-containing phases in the samples was done on the basis of hyperfine parameters, such as the hyperfine field, isomer shift, and quadrupole splitting, which were determined for the particular spectral components. Isomer shift values are related to the α-Fe standard. The conversion electron Mössbauer spectra measured as a function of annealing temperature of the coating are shown in Figure 12. The CEMS spectra revealed a distribution of Fe atoms in two magnetic and two paramagnetic Fe-Al environments in the annealed coatings.

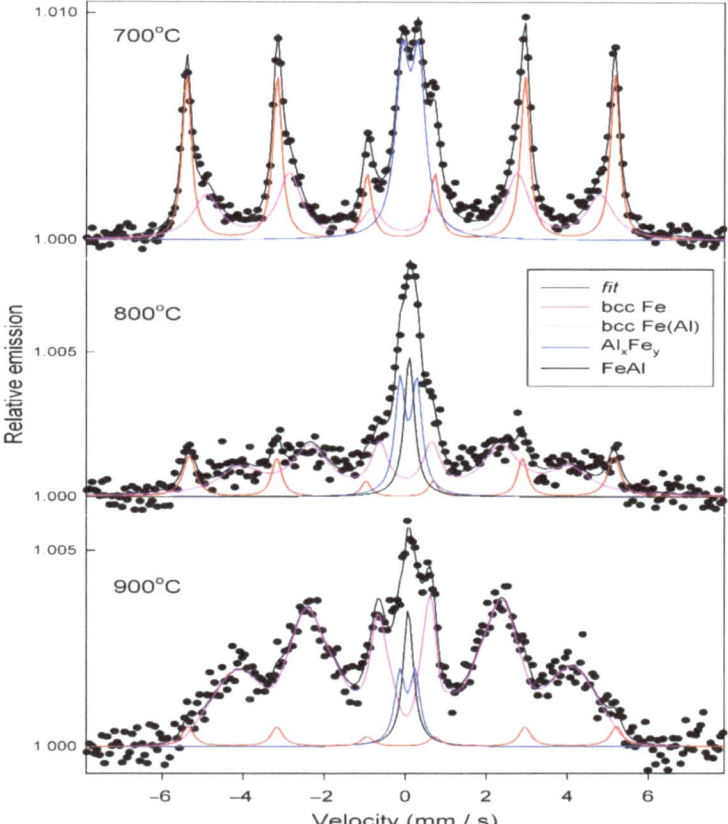

Figure 12. Conversion electron Mössbauer spectra of the coatings annealed at 700–900 °C.

The qualitative and quantitative analyses of the phase composition were performed based on fitting of the spectra with the use of the following spectral components:

- A magnetically split component (sextet) with the hyperfine field of 32.7 T, assigned to bcc Fe atomic environments without Al atoms as the nearest neighbors; however, some Al atoms were present in the remote vicinity of Fe atoms, thus causing a reduction of the hyperfine field of 32.95 T, characteristic for the pure bcc Fe phase [10,48,49];
- A sextet with broad lines and average hyperfine field values in the range of 25−30 T, originating from bcc Fe(Al) disordered solid solution;
- A quadrupole doublet with the quadrupole splitting of 0.40−0.44 mm/s and the isomer shift ranging from 0.19 to 0.23 mm/s, assigned to a paramagnetic Al-rich Al_xFe_y phase;
- A single line with the isomer shift of 0.22 mm/s, assigned to a paramagnetic intermetallic bcc FeAl phase.

The CEMS spectrum obtained after annealing at 700 °C was qualitatively similar to the spectrum of the untreated sample studied previously [10]. It consisted of the same three spectral components (1)–(3); however, the contribution of the quadrupole doublet (3) related to the formation of the Al-rich Al_xFe_y phase was markedly larger than for the unannealed sample. Annealing at 800–900 °C induced structural changes in the Fe-Al coating, which were clearly seen in the CEMS spectra as the intensity of the bcc Fe spectral component (1), dominating at 700 °C, which decreased significantly at higher temperatures in favor of the components related to binary Fe-Al phases. The broad sextet (2) revealed a distribution of the hyperfine field, particularly observed for the samples annealed at 800–900 °C. The hyperfine field values covered a wide range from about 10 T up to 33 T. The average values of the hyperfine field of the sextet (2) were approximately 30 T after annealing at 700 °C and 25 T after annealing at 800–900 °C.

This significant decrease of the average hyperfine field value with the increase of the annealing temperature indicates that a substantially larger number of Al atoms were incorporated into the Fe(Al) solid solution formed after annealing at 800–900 °C than at 700 °C. Based on the experimental dependence of the average hyperfine field on the composition of binary iron-rich Fe(Al) disordered solid solution, it was estimated that the percent of solute Al atoms increased from about 15% to about 28% after annealing at 700 °C and 800–900 °C, respectively [48,49].

Furthermore, a significant increase of the relative spectral fraction of the sextet (2) from 36% for the sample annealed at 700 °C to 56% and 83% after annealing at 800 °C and 900 °C, respectively, strongly indicates that the higher the annealing temperature, the more effective formation of the Fe(Al) solid solution.

Thermally induced changes in the magnetic Fe-Al environments were accompanied also by the evolution of paramagnetic components in the CEMS spectra. The quadrupole doublet observed after annealing at 700 °C was partially replaced by a single line after annealing at higher temperatures. As concerns the quadrupole doublet (3), its hyperfine parameters suggest the formation of an Al-rich phase with a composition close to Al_5Fe_2 [10,50], in good agreement with the XRD data. The appearance of the single line (4) after annealing at 800 and 900 °C strongly indicates the formation of cubic FeAl phase. The isomer shift of the single line was, however, considerably smaller than that characteristic of the ordered intermetallic FeAl phase with the equiatomic composition [49,50]. This fact suggests a non-stoichiometric ratio of the intermetallic phase, i.e., an excess of iron. The relative fraction of the intermetallic spectral component did not exceed 11%. Traces of paramagnetic iron oxides cannot be excluded.

The Mössbauer spectroscopy measurements showed that annealing induced further mixing of iron and aluminum in the samples so that the coating annealed at the highest temperature applied (900 °C) revealed the most homogenous structure, consisting predominantly of the bcc solid solution with an estimated average composition of $Fe_{72}Al_{28}$. The presence of a small amount of the intermetallic non-stoichiometric FeAl phase was also observed.

3.5. Hardness Analysis

Microhardness analysis has been conducted on the specimens for the cross-section of the coating/substrate system. The Vickers method was used (load 100 g for 10 s). Hardness analysis was realized by means of the Leitz–Wetzlar microhardness tester (LEICA, Wetzlar, Germany). In order to determine the uncertainty of measurement, t-student distribution was conducted, with a confidence level assumed at 95%. The obtained results were used to create a chart that presents the distribution of hardness.

The microhardness distribution presented in Figure 13 indicates significant differences in the hardness of the arc-spray-deposited coating before annealing and the coatings after the annealing processes, with three different temperature conditions described in heading 2.

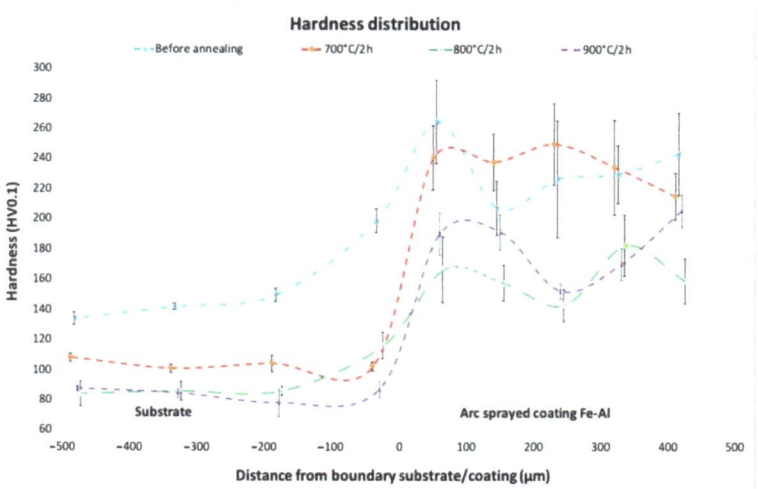

Figure 13. Distribution of microhardness in the cross-section of the substrate material and the arc-sprayed Fe-Al coating, before and after annealing at temperatures 700 °C, 800 °C, and 900 °C.

Annealing at a temperature of 700 °C did not significantly affect the hardness of the coating; it was comparable with the state before annealing. In addition, a hardness decrease in the steel substrate of approximately 40 HV0.1 was observed for all annealing treatments carried out. Annealing at a temperature of 800 °C caused a noticeable decrease in hardness in the coating by about 80 HV, as an additional effect a reduction in the standard deviation value was observed compared to the untreated coating and the coating annealed at 700 °C. The hardness of the annealed coating at 900 °C was between the hardness of the annealed coatings at 700 °C and 800 °C. In addition, a significant decrease in the value of the standard deviation from the average was noted (for 900 °C it was the lowest among the examined cases, Table 4), which indicates changes in the structure towards homogeneity.

Table 4. Hardness distribution in cross-section of substrate coating system, before and after annealing; the table shows the average value and standard deviation for the hardness distributions for the charts from Figure 12.

State of the Coating	Hardness Distribution, HV0.1								
	Approximately Distance from Boundary Substrate/Coating, µm								
	−470	−325	−180	−35	55	150	240	330	420
Before annealing	134 (7.91)	142 (4.3)	149 (8.32)	198 (15.99)	264.75 (55.36)	207.25 (36.31)	226.75 (77.82)	230 (38.22)	243.5 (54.55)
700 °C	108 (5.87)	100.83 (5.7)	104.05 (10.73)	102.3 (5.81)	240.75 (42.7)	237.75 (37.66)	249.75 (54.5)	234.5 (63.37)	215.25 (31.59)
800 °C	84.03 (16.46)	85.75 (12.53)	85.85 (5.85)	116.4 (17.82)	166.5 (43.87)	157.5 (23.69)	143 (21.44)	182.5 (40.6)	159 (29.67)
900 °C	87.35 (4.12)	84.33 (6.24)	77.83 (18.38)	86.68 (11.11)	190 (27.61)	191.25 (23.59)	152.25 (10.13)	170 (20.99)	205 (21.83)

4. Conclusions

The results of the SEM/EDX, XRD, CEMS, and HV0.1 experiments and analyses allowed for the evaluation of the changes in a chemical/phase composition and the degree of hardening in the arc-sprayed Fe-Al type coating after annealing at 700 °C, 800 °C, and 900 °C for 2 h.

It was determined that:

- The composite arc-sprayed Fe-Al coating with initial low participation of in situ created intermetallic phases showed significant changes in the phase composition, with an increase in the volume fraction of Fe-Al intermetallic phases as a result of annealing;
- At the lower annealing temperature of 700 °C, besides the Fe(Al) solid solution, other transitional Al-rich Fe2Al5 intermetallic phases were formed;
- An increase in the heating temperature induced further diffusion of Al and the formation of different Fe-Al-type phases, namely, bcc Fe(Al) solid solution and disordered intermetallic FexAly phases with varying chemical compositions (according to the Fe-Al equilibrium system);
- The thermal activation at 800 °C and 900 °C for 2 h stimulated the formation of the FeAl intermetallic phase, with specific {100} reflection originating from a superlattice with B2 ordering (as confirmed in XRD investigations);
- A significant decrease of the bcc Fe metallic phase (from the range of 50% to about 5%) in the arc-sprayed Fe-Al coating was observed, with the increase of annealing temperature up to 900 °C/2 h;
- The volume fraction of the B2 ordered FeAl phase increased with increasing annealing temperature;
- After annealing at the temperature of 900 °C, the structure was composed of a B2 ordered FeAl intermetallic phase and disordered Fe_3Al secondary solution, confirmed in the Mössbauer spectroscopy investigation;
- Heating of the arc-sprayed Fe-Al coating at a temperature of 900 °C for 2 h initiated the geometrical changes of lamellar structure, which ensured more homogeneity but was still not quite uniform in the SEM/EDX analysis;
- The microhardness distribution indicated significant differences in the hardness of the coatings after the annealing processes with three different temperature conditions;
- Annealing at temperatures of 800 °C and 900 °C caused a noticeable decrease in hardness;
- An additional effect was the reduction of the value of the standard deviation of the mean hardness value with the increase of the annealing temperature. The highest decrease in the value of the standard deviation from the mean hardness value occurred after annealing at the temperature of 900 °C, which confirmed the homogeneity changes of the structure.

Author Contributions: Conceptualization, T.C. and P.S.; methodology, M.C.; validation, B.S., M.C., and T.C.; investigation, A.P., A.G., and T.C.; resources, B.S.; data curation, B.S. and P.S.; writing—original draft preparation, T.C. and B.S.; writing—review and editing, T.C. and M.C., visualization, B.S.; supervision, T.C. and P.S.; designed and performed the experiments, M.C., P.S.; SEM, EDS spectroscopy, A.P.; Mössbauer spectroscopy, A.G. All authors have read and agreed to the published version of the manuscript.

Funding: The publication of the study was co-funded using the statutory subsidy of the Faculty of Production Engineering of the Warsaw University of Technology in 2019.

Institutional Review Board Statement: Not applicable.

Informed Consent Statement: Not applicable.

Data Availability Statement: Not applicable.

Acknowledgments: Thanks are given to the assistance from the SciTeeX company for arc spraying process.

Conflicts of Interest: The authors declare no conflict of interest.

References

1. Zamanzade, M.; Barnoush, A.; Motz, H. A review on the properties of iron aluminide intermetallics. *Crystals* **2016**, *6*, 10. [CrossRef]
2. Lisiecki, A.; Ślizak, D. Hybrid laser deposition of fe-based metallic powder under cryogenic conditions. *Metals* **2020**, *10*, 190. [CrossRef]
3. Ye, H.; Peng, S.; Yan, Z.; Zhang, X. Microstructure and properties of laser cladding Fe-Al intermetallics. *Adv. Mater. Res.* **2013**, *659*, 39–42. [CrossRef]
4. Wołczyński, W.; Senderowski, C.; Morgiel, J.; Garzeł, G. D-gun sprayed Fe-Al single particle solidification. *Arch. Metall. Mater.* **2014**, *59*, 211–220. [CrossRef]
5. Pawłowski, A.; Senderowski, C.; Wołczyński, W.; Morgiel, J. Major Detonation deposited Fe-Al coatings Part II: Transmission electron microscopy of interlayers and Fe-Al intermetallic coating detonation sprayed onto the 045 steel substrate. *Arch. Metall. Mater.* **2011**, *59*, 211–220. [CrossRef]
6. Pawłowski, A.; Czeppe, T.; Senderowski, C. Structure morphology of Fe-Al coating detonation sprayed onto carbon steel substrate. *Arch. Metall. Mater.* **2009**, *54*, 783–788.
7. Górka, J.; Czupryński, A.; Zuk, M.; Adamiak, M.; Kopyść, A. Properties and structure of deposited nanocrystalline coatings in relation to selected construction materials resistant to abrasive wear. *Materials* **2018**, *11*, 1184. [CrossRef] [PubMed]
8. Szczucka-Lasota, B.; Wegrzyn, T.; Stanik, Z.; Piwnik, J.; Sidun, P. Selected parameters of micro-jet cooling gases in hybrid spraying process. *Arch. Metall. Mater.* **2016**, *61*, 621–624. [CrossRef]
9. Czupryński, A.; Gorka, J.; Adamiak, M. Examining properties of arc sprayed nanostructured coatings. *Metalurgija* **2016**, *55*, 173–176.
10. Chmielewski, T.; Siwek, P.; Chmielewski, M.; Piątkowska, A.; Grabias, A.; Golański, D. Structure and selected properties of arc sprayed coatings containing in-situ fabricated Fe-Al intermetallic phases. *Metals* **2018**, *8*, 1059. [CrossRef]
11. Xu, B.; Zhu, Z.; Ma, S.; Zhang, W.; Liu, W. Sliding wear behavior of Fe-Al and Fe-Al/WC coatings prepared by high velocity arc spraying. *Wear* **2004**, *257*, 1089–1095. [CrossRef]
12. Cinca, N.; List, A.; Gärtner, F.; Guilemany, J.M. Influence of spraying parameters on cold gas spraying of iron aluminide intermetallics. *Surf. Coat. Technol.* **2015**, *268*, 99–107. [CrossRef]
13. Bober, M. Composite coatings deposited by plasma transfer—Characteristics and formation. *Weld. Int.* **2015**, *29*, 946–950. [CrossRef]
14. Cinca, N.; Guilemany, J.M. Thermal spraying of transition metal aluminides: An overview. *Intermetallics* **2012**, *24*, 60–72. [CrossRef]
15. Shishkovsky, I.V. Laser-controlled intermetallics synthesis during surface cladding. In *Laser Surface Engineering: Processes and Applications*; Woodhead Publishing: Sawston, UK, 2015; pp. 237–286. ISBN 9781782420798.
16. Nordmann, J.; Thiem, P.; Cinca, N.; Naumenko, K.; Kruger, M. Analysis of iron aluminide coated beams under creep conditions in high-temperature four-point bending tests. *J. Strain Anal. Eng. Des.* **2018**, *53*, 255–265. [CrossRef]
17. Doleker, K.M. The Examination of Microstructure and Thermal Oxidation Behavior of Laser-Remelted High-Velocity Oxygen Liquid Fuel Fe/Al Coating. *J. Mater. Eng. Perform.* **2020**, *29*, 3220–3232. [CrossRef]
18. Jozwik, P.; Bojar, Z.; Kołodziejczak, P. Microjoining of Ni3Al based intermetallic thin foils. *Mater. Sci. Technol.* **2010**, *26*, 473–477. [CrossRef]
19. Guilemany, J.M.; Cinca, N.; Fernández, J.; Sampath, S. Erosion, abrasive, and friction wear behavior of iron aluminide coatings sprayed by HVOF. *J. Therm. Spray Technol.* **2008**, *17*, 762–773. [CrossRef]
20. Szala, M.; Hejwowski, T. Zwiększanie odporności kawitacyjnej stopów metali przez napawanie powłok. *Weld. Technol. Rev.* **2015**, *87*, 57–60. [CrossRef]
21. Szala, M.; Walczak, M.; Pasierbiewicz, K.; Kamiński, M. Cavitation erosion and slidingwear mechanisms of AlTiN and TiAlN films deposited on stainless steel substrate. *Coatings* **2019**, *9*, 340. [CrossRef]
22. Chmielewski, T.; Hudycz, M.; Krajewski, A.; Salaciński, T.; Skowrońska, B.; Świercz, R. Structure investigation of titanium metallization coating deposited onto AlN ceramics substrate by means of friction surfacing process. *Coatings* **2019**, *9*, 845. [CrossRef]
23. Tomków, J.; Rogalski, G.; Fydrych, D.; Labanowski, J. Advantages of the application of the temper bead welding technique during wet welding. *Materials* **2019**, *16*, 915. [CrossRef]
24. Rajasekaran, R.; Lakshminarayanan, A.K.; Vasudevan, M.; Vasantharaja, P. Role of welding processes on microstructure and mechanical properties of nuclear grade stainless steel joints. *Proc. Inst. Mech. Eng. Part L J. Mater. Des. Appl.* **2019**, *233*, 2335–2351. [CrossRef]
25. Adamiak, M.; Górka, J.; Kik, T. Structure analysis of welded joints of wear resistant plate and constructional steel. *Arch. Mater. Sci. Eng.* **2010**, *46*, 108–114.
26. Tomków, J.; Fydrych, D.; Rogalski, G. Role of bead sequence in underwaterwelding. *Materials* **2019**, *12*, 3372. [CrossRef]
27. Wołosz, K.J.; Wernik, J. On the heat in the nozzle of the industrial pneumatic pulsator. *Acta Mech.* **2016**, *227*, 1111–1122. [CrossRef]
28. Senderowski, C.; Cinca, N.; Dosta, S.; Cano, I.G.; Guilemany, J.M. The Effect of Hot Treatment on Composition and Microstructure of HVOF Iron Aluminide Coatings in Na_2SO_4 Molten Salts. *J. Therm. Spray Technol.* **2019**, *28*, 1492–1510. [CrossRef]

29. Wernik, J.; Wolosz, K.J. Study of heat transfer in fins of pneumatic pulsator using thermal imaging. *Chem. Eng. Trans.* **2015**, *45*, 985–990. [CrossRef]
30. Hodulova, E.; Ramos, A.S.; Kolenak, R.; Kostolny, I.; Simekova, B.; Kovarikova, I. Characterization of ultrasonic soldering of Ti and Ni with Ni/Al reactive multilayer deposition. *Weld. Technol. Rev.* **2019**, *91*, 51–57. [CrossRef]
31. Hong, L.; Xuan, L.; Haixin, H. Microstructure and properties of ZrO2 ceramic and Ti-6A-4V alloy vacuum brazed by Ti-28Ni filler metal. *Weld. Technol. Rev.* **2019**, *91*, 35–41. [CrossRef]
32. Shishkovsky, I.; Missemer, F.; Kakovkina, N.; Smurov, I. Intermetallics synthesis in the Fe-Al system via layer by layer 3D laser cladding. *Crystals* **2013**, *3*, 517–529. [CrossRef]
33. Sun, K.; Cheng, J.; Liu, X.; Fang, L.; Ma, L. In-Situ Fabrication of Fe–Al Intermetallic Coating by Laser Remelting. *J. Mechatron.* **2014**. [CrossRef]
34. Ma, H.; Ren, K.; Xiao, X.; Qiu, R.; Shi, H. Growth characterization of intermetallic compounds at the Cu/Al solid state interface. *Mater. Res. Express* **2019**, *6*, 1–11. [CrossRef]
35. Chu, Y.J.; Li, X.Q.; Li, J.; Yang, F.; Yang, S.F. Effects of annealing temperature on microstructure and properties of AlFeCrCoNiTi high-entropy alloy coating prepared by laser cladding. *Cailiao Rechuli Xuebao/Trans. Mater. Heat Treat.* **2018**. [CrossRef]
36. Kik, T.; Moravec, J.; Novakova, I. New method of processing heat treatment experiments with numerical simulation support. In Proceedings of the IOP Conference Series: Materials Science and Engineering, Hong Kong, China, 12–14 December 2017; p. 227.
37. Winczek, J.; Gawronska, E.; Gucwa, M.; Sczygiol, N. Theoretical and experimental investigation of temperature and phase transformation during SAW overlaying. *Appl. Sci.* **2019**, *9*, 1472. [CrossRef]
38. Deevi, S.C.; Sikka, V.K. Nickel and iron aluminides: An overview on properties, processing, and applications. *Intermetallics* **1996**, *4*, 357–375. [CrossRef]
39. Sun, Y.; Zhang, Z.; Jin, X.; Xu, B.; Zhao, G. Cutting force models for Fe–Al-based coating processed by a compound NC machine tool. *Int. J. Adv. Manuf. Technol.* **2015**, *79*, 693–704. [CrossRef]
40. Cegan, T.; Petlak, D.; Skotnicova, K.; Jurica, J.; Smetana, B.; Zla, S. Metallurgical preparation of Nb-Al and W-Al intermetallic compounds and characterization of their microstructure and phase transformations by DTA technique. *Molecules* **2020**, *25*, 2001. [CrossRef]
41. Spyra, J.; Michalak, M.; Niemiec, A.; Łatka, L.A. Mechanical properties investigations of the plasma sprayed coatings based on alumina powder. *Weld. Technol. Rev.* **2020**, *92*, 17–23. [CrossRef]
42. Dean, S.W.; Potter, J.K.; Yetter, R.A.; Eden, T.J.; Champagne, V.; Trexler, M. Energetic intermetallic materials formed by cold spray. *Intermetallics* **2013**, *43*, 121–130. [CrossRef]
43. Novák, P.; Michalcová, A.; Marek, I.; Mudrová, M.; Saksl, K.; Bednarčík, J.; Zikmund, P.; Vojtěch, D. On the formation of intermetallics in Fe-Al system—An in situ XRD study. *Intermetallics* **2013**, *32*, 127–136. [CrossRef]
44. Wang, H.T.; Li, C.J.; Yang, G.J.; Li, C.X. Cold spraying of Fe/Al powder mixture: Coating characteristics and influence of heat treatment on the phase structure. *Appl. Surf. Sci.* **2008**, *255*, 2538–2544. [CrossRef]
45. Naoi, D.; Kajihara, M. Growth behavior of Fe_2Al_5 during reactive diffusion between Fe and Al at solid-state temperatures. *Mater. Sci. Eng. A* **2007**, *459*, 375–382. [CrossRef]
46. Pawlowski, L. Thermal Spraying Techniques. In *The Science and Engineering of Thermal Spray Coatings*; John Wiley & Sons: Hoboken, NJ, USA, 2008; pp. 67–113.
47. Michalak, M.; Łatka, L.; Sokołowski, P.; Niemiec, A.; Ambroziak, A. The microstructure and selected mechanical properties of Al_2O_3 + 13 wt % TiO_2 plasma sprayed coatings. *Coatings* **2020**, *10*, 173. [CrossRef]
48. Perez Alcazar, G.A.; Galvao Da Silva, E. Mossbauer effect study of magnetic properties of Fe1-qAl q, 0. *J. Phys. F Met. Phys.* **1987**, *17*, 2323–2335. [CrossRef]
49. Krasnowski, M.; Grabias, A.; Kulik, T. Phase transformations during mechanical alloying of Fe-50% Al and subsequent heating of the milling product. *J. Alloys Compd.* **2006**, *424*, 119–127. [CrossRef]
50. Nasu, S.; Gonser, U.; Preston, R.S. Defects and phases of iron in aluminium. *J. Phys. Colloq.* **1980**, *41*, 385–386. [CrossRef]

Effect of APS Spraying Parameters on the Microstructure Formation of Fe₃Al Intermetallics Coatings Using Mechanochemically Synthesized Nanocrystalline Fe-Al Powders

Cezary Senderowski [1,*], Nataliia Vigilianska [2], Oleksii Burlachenko [2], Oleksandr Grishchenko [2], Anatolii Murashov [2] and Sergiy Stepanyuk [3]

[1] Institute of Mechanics and Printing, Faculty of Mechanical and Industrial Engineering, Warsaw University of Technology, 02-524 Warsaw, Poland
[2] Department of Protective Coatings, E.O. Paton Electric Welding Institute, 03680 Kiev, Ukraine
[3] Department of Studies of Physical-Chemical Processes in Welding Arc, E.O. Paton Electric Welding Institute, 03680 Kiev, Ukraine
* Correspondence: cezary.senderowski@pw.edu.pl

Abstract: The present paper presents a study of the behaviour of Fe₃Al intermetallic powders particles based on 86Fe-14Al, 86Fe-14(Fe5Mg), and 60.8Fe-39.2(Ti37.5Al) compositions obtained by mechanochemical synthesis at successive stages of the plasma spraying process: during transfer in the volume of the gas stream and deformation at the moment of impact on the substrate. The effect of the change in current on the size of powder particles during their transfer through the high-temperature stream and the degree of particle deformation upon impact with the substrate was determined. It was found that during transfer through the plasma jet, there was an increase in the average size of sputtering products by two–three times compared to the initial effects of mechanochemical synthesis due to the coagulation of some particles. In this case, an increase in current from 400 to 500 A led to a growth in average particle size by 14–47% due to the partial evaporation of fine particles with an increase in their heating degree. An increase in current also led to a 5–10% growth in particle deformation degree upon impact on the substrate due to the rising temperature and velocity of the plasma jet. Based on the research, the parameters of plasma spraying of mechanically synthesized Fe₃Al intermetallic-based powders were determined, at which dense coatings with a thin-lamellar structure were formed.

Keywords: plasma spraying; intermetallic coating; mechanochemical synthesis; conglomerate; splat test; deformation degree

1. Introduction

Iron aluminide-based intermetallic alloys, due to their physicochemical and mechanical properties, as well as their stable structure and resistance to high-temperature corrosion, are promising materials for heat exchangers, nuclear reactor components, automotive exhaust systems, etc. [1–5]. The main advantage of iron aluminides over heat-resistant nickel alloys and stainless steels is the availability and low cost of the base iron component, as well as their ease of processing.

Iron aluminides have found a wide practical application as protective coatings obtained by thermal spraying, especially HVOF, plasma (PS), and Arc and D-gun spraying [2,5–11]. Unlike in D-gun spraying [10], during the PS process, powder particles falling into the high-temperature plasma jet (PJ) are melted and transferred to the substrate as droplets. In high-temperature flight, phenomena such as dispersion, coagulation, and changes in microstructure and phase composition occur in particles [12,13]. When the molten drop impacts the surface of the substrate, it spreads, solidifies, and forms a surface layer in the form of a splat. The structure and properties of coatings depend on the powder particles' state during PS spraying. Completely molten particles contribute to the formation of dense layers and lead to reduced porosity. The degree of particle melting depends on the characteristics of the PJ (velocity and temperature, viscosity and thermal conductivity of the gas environment, degree of dissociation and ionisation of the gas molecules) and the material properties of the atomised particles (density, heat capacity, thermal conductivity, heat of fusion) [14].

At present, one of the ways to obtain composite intermetallic powders of the Fe-Al system for thermal spraying is the mechanochemical synthesis method (MCS) [15–17].

It should be noted, however, that unlike the D-gun process [9,10], the literature needs a comprehensive approach to studying the formation of plasma coatings using intermetallic powders of the Fe-Al system, including those obtained by the MCS method. Therefore, this study aimed to investigate the effect of changing the current intensity on the physicochemical processes occurring during forming Fe_3Al-based intermetallic coatings by plasma spraying.

2. Materials and Methods

Iron aluminide powders obtained by mechanochemical synthesis were used as research material, while iron, aluminium, aluminium alloy (Al5Mg), and Ti37.5Al intermetallic powders were applied as starting materials. Using Ti as an alloying element allows for realising several mechanisms of iron intermetallic strengthening, namely structure ordering, strengthening with dispersed inclusions, and the formation of coherent microstructures. Titanium differs by significant solubility in a solid state in Fe–Al phases that result in Fe_3Al structure stabilising around the FeAl structure at high temperatures. Strengthening with dispersed precipitations of hexagonal Laves phase $(Fe, Al)_2Ti$ can take place in addition to strengthening due to structure ordering in the Fe–Al–Ti system. Furthermore, there is a specific range of composition in the Fe–Al–Ti system, where coherent structures [18] are formed. If an element such as Mg is used for alloying powders based on the Fe_3Al intermetallic compound, it is possible to expect strengthening by incoherent compounds. Commercially available powders of the alloys Al5Mg and Ti37.5Al were used instead of the pure elements of aluminium, magnesium, and titanium to reduce the degree of oxidation of the MCS products.

The MCS process was performed in a planetary-type mill «Aktivator 2SL» for 5 h. The relation of the mass of balls to the mass of powder was 10:1. The central axis of the mill triboreactor was rotated at a 100 rpm rate; drums rotated around their axis at a 1500 rpm rate. Parts of the jar and milling agents were manufactured of 100Cr6 steel. The MCS process was performed in the air. Surface-active substances (SAS), namely oleic acid, were added to the mixture to prevent pickup of processed charge on the milling agents and jar wall, and to intensify the process of new synthesis phases.

The amount of aluminium alloy powder introduced in the mixture with iron powder was selected for formation in MCS of $Fe_3(Al, Mg)$ intermetallics in the case of AlMg that corresponded to 14 wt.% of Al-alloy and $(Fe, Ti)_3Al$ in the case of TiAl intermetallics. In the latter, the variant amount of introduced TiAl was 39.2 wt.%.

The microstructure, chemical, and phase composition of the MCS powder products (as an intermetallic compound based on Fe_3Al iron aluminide) were confirmed, respectively, by SEM/EDS and XRD analysis using a JEOL 5310 microscope (Japan) operating at 20 kV, and a XRD Simens D-500 diffractometer (Germany) with $CoK\alpha$ radiation ($\lambda = 0.178897$ nm). An angular step size of $0.02°$/min and a step time of 5 s per point were used, respectively.

The appearance and X-ray patterns of MCS powders of Fe_3Al, Fe-AlMg, and Fe-TiAl systems are shown in Figures 1 and 2. The chemical composition of MCS powders is presented in Table 1.

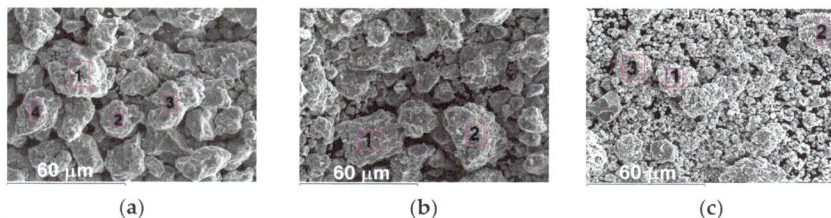

Figure 1. Appearance MCS powders: (**a**) Fe$_3$Al, (**b**) Fe-AlMg, (**c**) Fe-TiAl.

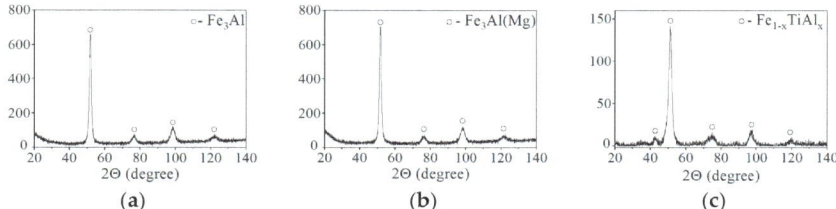

Figure 2. X-ray patterns of MCS powders: (**a**) Fe$_3$Al, (**b**) Fe-AlMg, (**c**) Fe-TiAl.

Table 1. Chemical composition (wt.%) of MCS powders.

Spectrum	Fe	Al	Ti	Mg	O
Fe$_3$Al Figure 1a					
1	79.22	16.91	-	-	3.87
2	81.78	14.04	-	-	4.18
3	81.35	13.51	-	-	5.14
4	80.15	14.79	-	-	5.06
Fe-AlMg Figure 1b					
1	82.14	13.99	-	0.62	3.25
2	82.49	12.39	-	0.55	4.57
Fe-TiAl Figure 1c					
1	61.72	12.03	21.21	-	5.04
2	59.43	13.59	22.17	-	4.81
3	60.97	11.14	24.24	-	3.65

The characteristics of the MCS powders used in work for investigating the coating formation process during plasma spraying are shown in Table 2 [15,16].

Table 2. Characteristics of Fe$_3$Al–based intermetallic MCS powders.

System	Composition, wt.%	Phase Composition	Crystallite Size, nm	Microhardness, HV0.01 MPa	Particle Size, μm		
					D_{10}	D_{50}	D_{90}
Fe$_3$Al	86Fe+14Al	Fe$_3$Al	15	4060 ± 1010	3.6	11.2	32.9
Fe-AlMg	86Fe+14(Al5Mg)	solid solution Mg in Fe$_3$Al (Fe$_3$Al(Mg))	14	4630 ± 950	2.8	14.5	29.8
Fe-TiAl	60.8Fe+39.2(Ti37.5Al)	solid solution Al in FeTi (Fe$_{1-x}$TiAl$_x$)	10	3400 ± 1120	2.6	8.7	29.7

The crystallite size was estimated using the Scherrer equation:

$$D = k\lambda / \beta \cos\theta$$

where k is the Scherrer constant (≈0.94), λ is the wavelength of the radiation used (for Co λ = 1.78897 Å), θ is the reflection angle, β is the true broadening of the X-ray line.

Based on the crystallite size calculations, it can be noted that the powders obtained by the MCS method are nanocrystalline.

The absence of fluidity in MCS products is related to the high specific surface area of the particles, and the challenge is to uniformly feed these powders into the stream during PS [19]. To uniformly feed MCS products into PJ, they were conglomerated with a 5% polyvinyl alcohol solution in water, dried, and sifted for 40–80 μm particles.

The choice of PS modes for Fe_3Al–based intermetallic materials was carried out using the CASPSP software version 3.1 [20]. This software is designed for the computer simulation of turbulent plasma jets used in coating spraying and for modelling the heating and transport of atomised particles in such jets. Based on the analysis of the heating and transport of Fe_3Al particles in the range of 40–80 μm in PJ, it was found that, in terms of total melting and the lack of evaporation of the particles, the most reasonable parameters are a current of 400–500 A and a plasma gas (PG) $Ar+N_2$ flow rate of 25 SLPM (Figure 3). The use of other modes for spraying Fe_3Al-based powders is irrational due to the absence of the complete melting of particles at a current of less than 400 A (Figure 4a) and the possibility of the material evaporation at a current of more than 500 A (Figure 4b).

Powder spraying was carried out using the UPU–8M atmospheric plasma spraying device at different arc current (I) parameters. Changing the current significantly affects the temperature and velocity of PJ, which determines the heating and velocity of the particles during transport in the stream [21]. The voltage, plasma gas (mixture $Ar+N_2$) flow rate, and powder feed rate were constant in all experiments. The Ar/N_2 ratio (7.3/1) and the flow rate of the plasma gas mixture (25 SLPM) made it possible to ensure the stable operation of the plasmatron at a voltage of 40 V. The modes of PS operation are presented in Table 3.

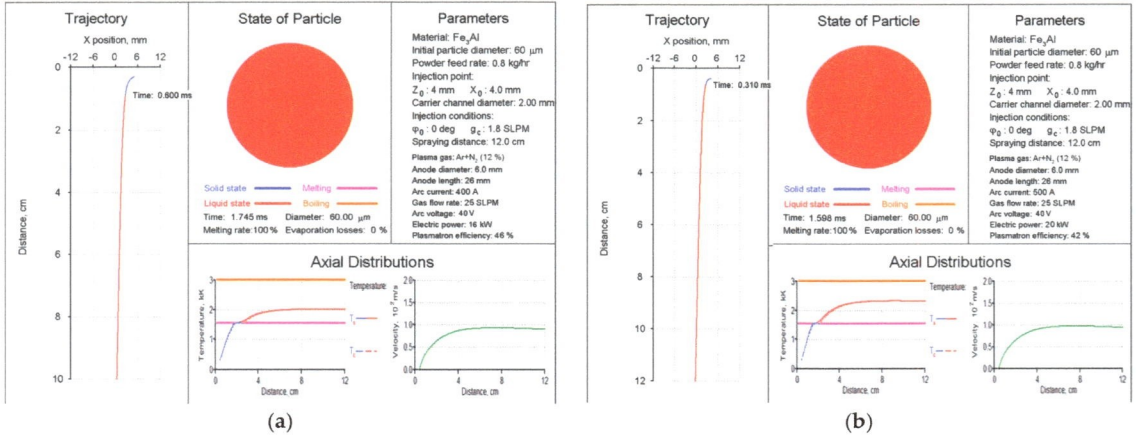

Figure 3. Graphical result of simulation by the CASPSP software of plasma spraying of Fe_3Al particles at the current of 400 (**a**) and 500 A (**b**).

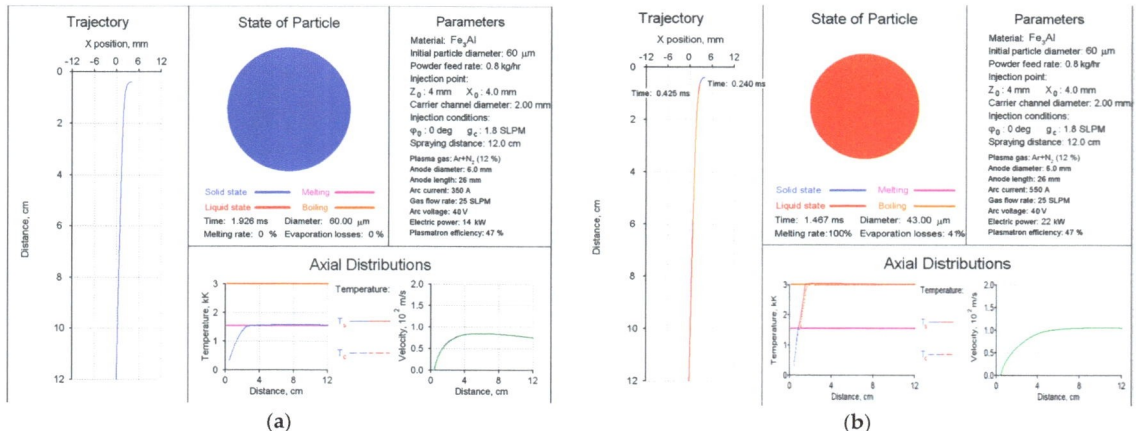

Figure 4. Graphical result of simulation by the CASPSP software of plasma spraying of Fe_3Al particles at the current of 350 (**a**) and 550 A (**b**).

Table 3. Plasma spray modes for spraying of Fe_3Al–based intermetallic particles.

Modes No.	Current, A	Voltage, V	Power, kW	PG Flow Rate, SLPM		Powder Feed Rate, g/min
				Ar	N₂	
1	400		16			
2	450	40	18	22	3	12
3	500		20			

Studies of the PS processes occurring in the powder during spraying were carried out by collecting particles transferred through the plasma jet into the water at a distance of 120 mm from the end of the plasma torch to the water's surface. The analysis of the size and microstructure of the particles of sprayed powders collected in the water bath will determine the processes taking place with the powder particles in a plasma jet, on which the further formation of the coating structure depends.

The state of the sprayed particle material after impact on the substrate was studied using the splat test. Spraying was carried out by moving polished stainless steel plates of 50 × 30 × 0.5 mm in a plane perpendicular to the jet axis (Figure 5). As a result, single particles of sprayed material deformed upon contact with the substrate surface (splats) were fixed on the samples.

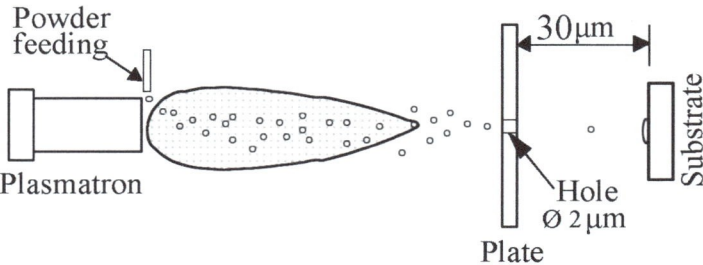

Figure 5. Scheme of the splat test.

To investigate powders sprayed into the water and evaluate the degree of deformation of particles after their impact on the substrate, a complex technique was used, including the analysis of particle size distribution–ASOD-300 laser analyser (Novatek-Electro, Odesa, Ukraine); metallographic examinations–optical microscope "Neofot-32" with an attachment for digital photography; measurement of splats diameters (D) and classification of splats by appearance–optical microscope Jenavert; scanning electron microscopy (SEM)–scanning electron microscope JSM-6390LV (JEOL, Warsaw, Poland) with an attachment for energy dispersive analysis INCA in the secondary electron mode, in low vacuum (10^{-4} Pa), with an accelerating voltage of 20 kV.

Chemical etching of metallographic cross-sections was used to reveal the structure of powders sprayed into water. For Fe_3Al and Fe-AlMg powders, a 10% alcoholic solution of nitric acid was used for 4–5 min; the Fe-TiAl powder was etched with a solution of HF+HCl+HNO_3+water for 2–3 min. Etching was carried out for the powders sprayed in Mode 3.

The microhardness of the powders and coatings was determined in a microhardness tester PMT-3. To quantitatively analyse the content of pores in the coatings, an optical technique (image analysis method) was used, which consists of determining the area per detected pores relative to the entire area of the coating cross-section (ASTM E2109-01). Image-Pro Plus 7 software processed the digital image of the microstructure of the coatings.

3. Results and Discussion

The appearance of powders of Fe_3Al, Fe-AlMg, and Fe-TiAl systems sprayed into the water at different currents is shown in Figures 6–8. Analysis of the appearance of the powders showed that most of the particles (~95%) are spherical, indicating their complete melting during transferring through the plasma jet.

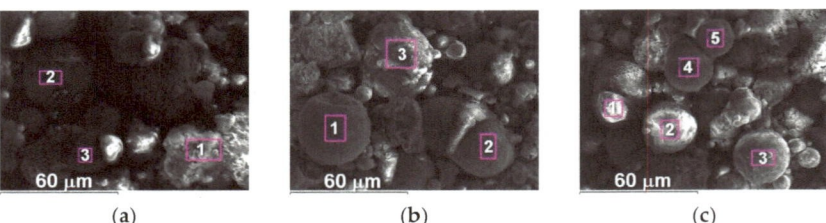

Figure 6. The appearance of Fe_3Al powders after transfer through the plasma jet: (**a**) I = 400 A, (**b**) I = 450 A, (**c**) I = 500 A.

Figure 7. The appearance of Fe-AlMg powders after transfer through the plasma jet: (**a**) I = 400 A, (**b**) I = 450 A, (**c**) I = 500 A.

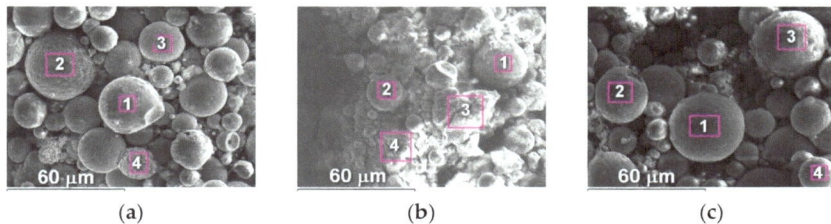

Figure 8. The appearance of Fe-TiAl powders after transfer through the plasma jet: (**a**) I = 400 A, (**b**) I = 450 A, (**c**) I = 500 A.

It is known [22] that when a particle containing aluminium enters the oxygen-containing zones in a plasma stream, the process of aluminium oxidation develops with the appearance of an aluminium oxide film on the particle surface. By analysing the chemical composition of the powders, the presence of oxygen on the particle surfaces can be noted (Tables 4–6), which indicates the formation of oxides (in particular, aluminium oxide–Figure 6c, Table 4, Spectrum 1; Figure 7b, Table 5, Spectrum 1).

Table 4. Chemical composition (wt.%) of Fe_3Al powder particles after transfer through the plasma jet.

Figure	Spectrum	Fe	Al	O
Figure 6a	1	57.48	5.6	36.92
	2	84.31	9.96	5.73
	3	61.59	11.49	26.92
Figure 6b	1	76.83	10.08	13.09
	2	86.18	5.86	7.96
	3	36.36	30.7	32.94
Figure 6c	1	2.85	48.40	48.75
	2	82.97	6.62	10.41
	3	62.38	11.20	26.42
	4	76.40	10.48	13.12
	5	70.22	12.27	17.51

Table 5. Chemical composition (wt.%) of Fe-AlMg powder particles after transfer through the plasma jet.

Figure	Spectrum	Fe	Al	Mg	O
Figure 7a	1	77.78	6.64	0.27	15.31
	2	71.58	5.76	0.52	22.14
	3	59.43	17.15	0.85	22.57
Figure 7b	1	9.56	48.93	0.30	41.21
	2	60.76	15.65	1.04	22.55
Figure 7c	1	61.89	11.2	0.81	26.1
	2	75.44	10.32	0.58	13.66
	3	52.52	10.55	0.3	36.63
	4	86.01	6.26	0.26	7.47

Table 6. Chemical composition (wt.%) of Fe-TiAl powder particles after transfer through the plasma jet.

Figure	Spectrum	Fe	Al	Ti	O
Figure 8a	1	15.63	13.9	17.36	53.11
	2	20.59	12.92	17.78	48.71
	3	12.33	21.65	8.49	57.53
	4	16.41	14.32	15.6	53.67
Figure 8b	1	19.89	10.96	23.7	45.45
	2	3.87	6.0	45.85	44.28
	3	60.86	3.7	3.4	32.04
	4	24.5	9.09	19.4	47.01
Figure 8c	1	9.36	19.71	27.62	43.31
	2	19.16	11.92	32.08	36.84
	3	8.04	13.26	30.9	47.80
	4	6.32	7.46	40.47	45.75

The oxygen content in powders of the Fe-TiAl system is, on average, two times higher than the oxygen content in Fe_3Al and Fe-AlMg powders. This is explained by the tendency of Ti and Al powder components to oxidize and their higher content in the initial powder mixtures. The reduced iron content in powders of the Fe-TiAl system also indicates the formation of an oxide film on the particle surface.

The microstructure of powders (Figures 9–11) shows that oxides on the surface of the particles are arranged in the form of thin films or obtain a domed shape as a result of the movement of particles in the turbulent plasma stream (indicated by arrows at Figures 9, 10 and 11c).

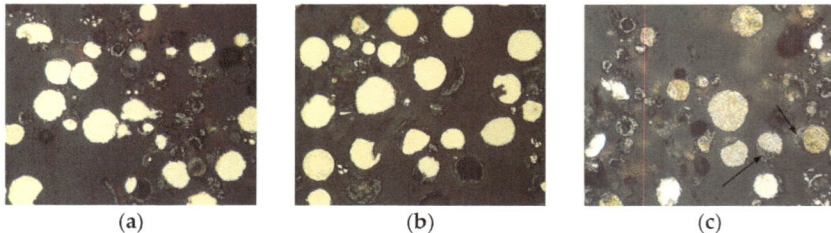

Figure 9. Microstructure (×400) of Fe_3Al powders after transfer through the plasma jet: (**a**) I = 400 A, (**b**) I = 450 A, (**c**) I = 500 A (etched).

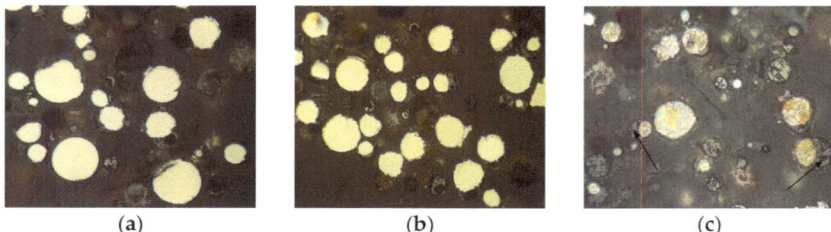

Figure 10. Microstructure (×400) of Fe-AlMg powders after transfer through the plasma jet: (**a**) I = 400 A, (**b**) I = 450 A, (**c**) I = 500 A (etched).

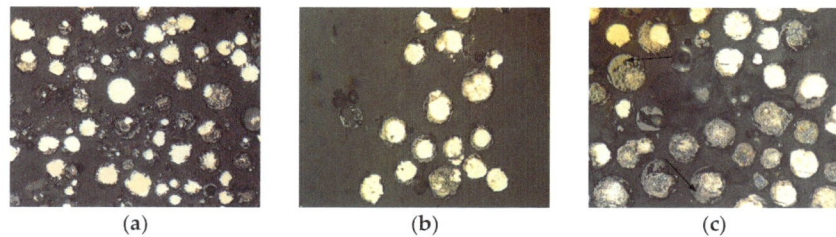

Figure 11. Microstructure (×400) of Fe-TiAl powders after transfer through the plasma jet: (**a**) I = 400 A, (**b**) I = 450 A, (**c**) I = 500 A (etched).

Histograms of the particle size distribution of Fe$_3$Al, Fe-AlMg, and Fe-TiAl powders after the conglomerates transferred through the plasma stream at different currents are shown in Figures 12–14. It can be seen that most of the powder particles (>55%) are in the size range of 10...30 µm. The reduction in particle size compared to the conglomerates fed into the plasma stream is related to the binder burnout during the interaction of the high-temperature plasma stream with the conglomerates.

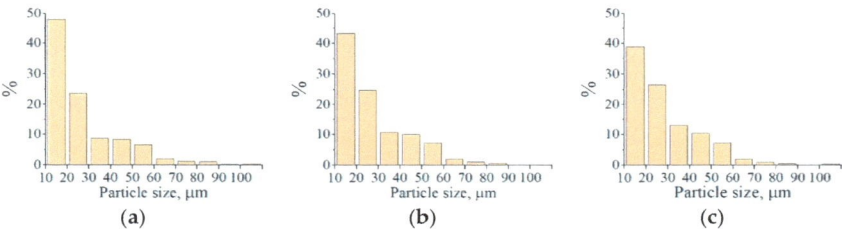

Figure 12. Histograms of the particle size distribution of Fe$_3$Al powders sprayed into the water: (**a**) I = 400 A, (**b**) I = 450 A, (**c**) I = 500 A.

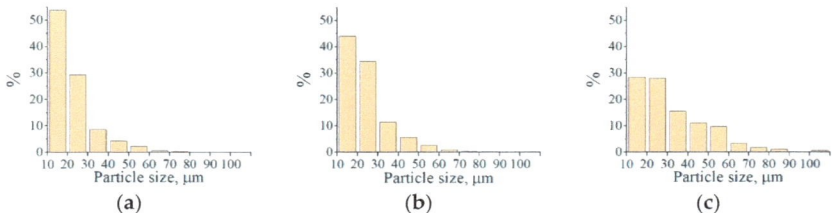

Figure 13. Histograms of the particle size distribution of Fe-AlMg powders sprayed into the water: (**a**) I = 400 A, (**b**) I = 450 A, (**c**) I = 500 A.

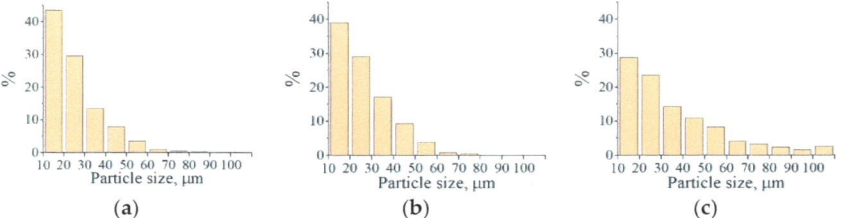

Figure 14. Histograms of the particle size distribution of Fe-TiAl powders sprayed into the water: (**a**) I = 400 A, (**b**) I = 450 A, (**c**) I = 500 A.

MCS products also simultaneously fuse with the destruction of conglomerates in the plasma stream. As a result, the average particle size increases 2–3 times relative to the initial MCS products (Table 7).

Table 7. Granulometric composition of powders after transfer through the plasma jet.

Powders	Current, A	Particle Size, μm		
		D_{10}	D_{50}	D_{90}
Fe$_3$Al	400	12	22	52
	450	13	25	52
	500	13	25	52
Fe-AlMg	400	13	19	40
	450	13	22	43
	500	13	28	58
Fe-TiAl	400	13	22	43
	450	13	25	46
	500	13	31	70

The increase in current leads to a 38–57% decrease in the number of particles within the size range of 10...20 μm and a 14–47% increase in the average particle size (Figure 15). This may be due to the partial evaporation of small particles with increased heating.

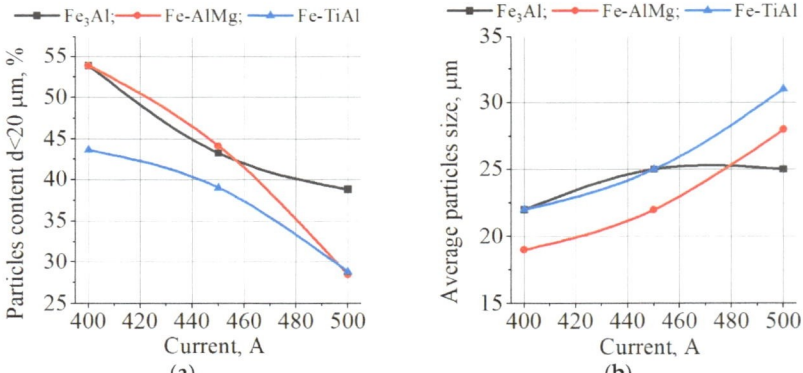

Figure 15. Effect of current change: (a) on particle content of <20 μm; (b) on the average particle size of powders.

Analysis of the shape of individual splats on substrates is one of the factors in optimizing plasma spraying methods. The degree of particle deformation D/d on impact with the sprayed surface determines the particle's surface contact area. The larger deformed particle size D compared to the initial particle size d in the gas stream in front of the sprayed surface, the greater probability of strong adhesion of the contacting materials, with all other conditions being equal.

The appearance of splashes obtained by spraying Fe$_3$Al, Fe-AlMg, and Fe-TiAl powders as a function of current intensity is shown in Figures 16–18.

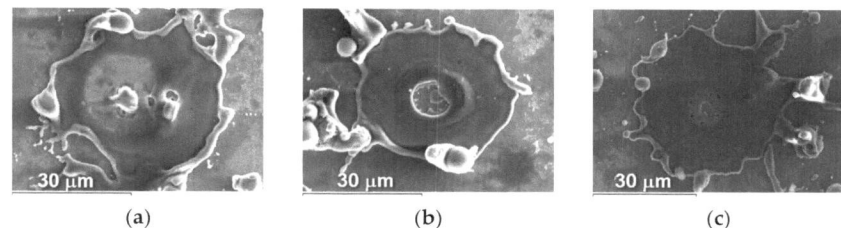

Figure 16. Splashes of Fe$_3$Al powders after impact with steel substrate at (**a**) I = 400 A; (**b**) I = 450 A; (**c**) I = 500 A.

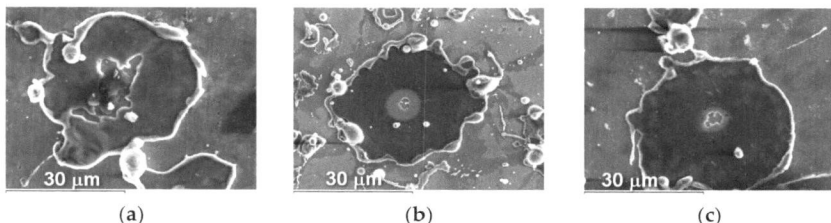

Figure 17. Splashes of Fe-AlMg powders after impact with steel substrate at (**a**) I = 400 A; (**b**) I = 450 A; (**c**) I = 500 A.

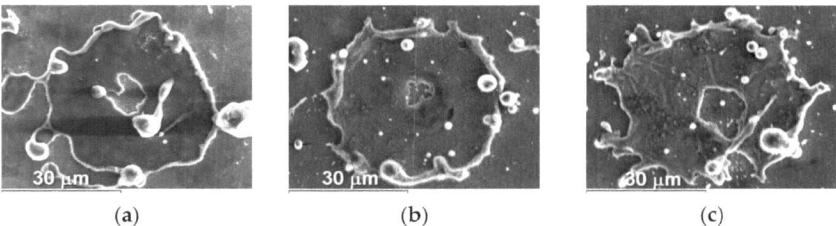

Figure 18. Splashes of Fe-TiAl powders after impact with steel substrate at (**a**) I = 400 A, (**b**) I = 450 A, (**c**) I = 500 A.

From the splats obtained during the interaction of powder particles with the substrate in all three modes, it can be concluded that the particles are in a fully molten state at the moment of impact with the substrate and have a disk shape. After the impact of the particles on the substrate and spread across the surface, the central part of the splats turns out to be unfilled by the material. This is explained by the fact that, inside the droplet, cavitation processes occur when it hits the surface of a solid; i.e., bubbles form and grow as the pressure drops to the saturation vapour pressure. The bubbles break through the liquid coating of the droplet and form crater-like holes in the deformed powder particle [23].

The results of estimating the degree of deformation of D/d particles after impact on the ground are shown in Table 8.

The dependence of the average splat diameter and the degree of particle deformation on the current is shown in Figure 19. As can be seen, as the current increases, the average diameter of the splats and the degree of deformation of the particles also increase. This is due to the increase in the temperature and PJ velocity, which leads to a decrease in the surface tension and viscosity of the molten particle, and an increase in the impulse and pressure that act on the particle when it impacts the substrate.

Table 8. Degree of deformation of D/d particles after impact on the substrate.

Current, A	Powders	Average Particles Size d, μm	Average Splats Size D, μm	Particles Deformation Degree D/d
400	Fe₃Al	22	46	2.1
	Fe-AlMg	19	49	2.6
	Fe-TiAl	22	42	1.9
450	Fe₃Al	25	55	2.2
	Fe-AlMg	22	57	2.6
	Fe-TiAl	25	53	2.1
500	Fe₃Al	25	55	2.2
	Fe-AlMg	28	78	2.8
	Fe-TiAl	31	65	2.1

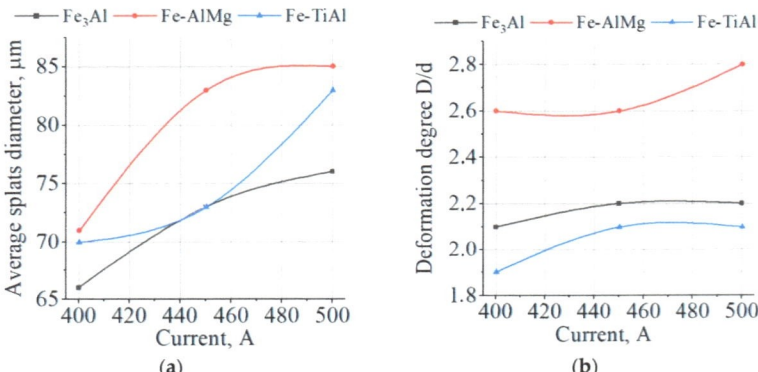

Figure 19. Effect of current change on (**a**) average splats' diameter; (**b**) particles' deformation degree.

As a result of the plasma spraying of these powders, the coating layers are formed from completely melted and deformed particles, so the coatings have a dense thin-lamellar structure with a small number of oxide films at the lamella boundaries (Figures 20–22). The coatings have dense adherence to the steel substrate, and no delamination defects are observed.

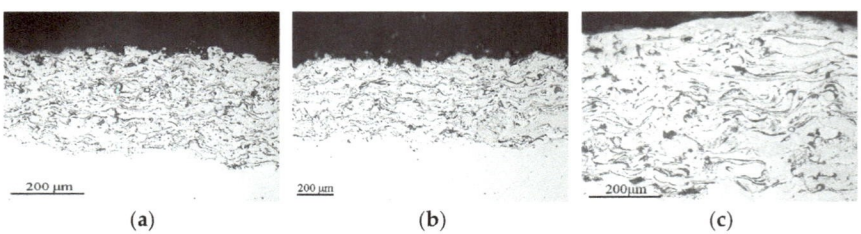

Figure 20. Microstructure of Fe₃Al plasma coatings sprayed at: (**a**) I = 500 A; (**b**) I = 450 A; (**c**) I = 500 A.

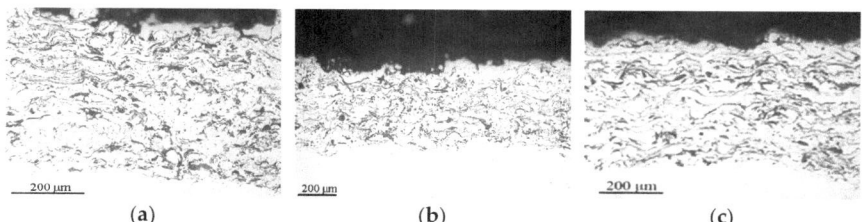

Figure 21. Microstructure of Fe-AlMg plasma coatings sprayed at (**a**) I = 500 A; (**b**) I = 450 A; (**c**) I = 500 A.

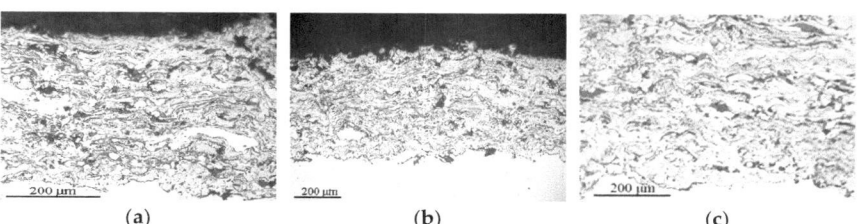

Figure 22. Microstructure of Fe-TiAl plasma coatings sprayed at (**a**) I = 500 A; (**b**) I = 450 A; (**c**) I = 500 A.

During the spraying of coatings of intermetallic powders, their phase composition does not completely coincide with the composition of the initial powders, which is associated with the active development of the process of particle oxidation during their flight. Oxides are present in all coatings. In the Fe$_3$Al coating, Al$_2$O$_3$ aluminium oxide is present; in the Fe-AlMg coating, complex oxide MgAl$_2$O$_4$ and in the Fe-TiAl coating, titanium oxide TiO are present (Figure 23). In the Fe-AlMg coating, as in the original MCS powder, a solid solution of Mg in the Fe$_3$Al intermetallic compound is retained. In the Fe$_3$Al coating, the FeAl intermetallic phase is noted in addition to the base phase. During the spraying of the Fe-TiAl powder, the solid solution (Fe$_{1-x}$TiAl$_x$) was transformed into the intermetallic phase (Fe, Ti)$_3$Al.

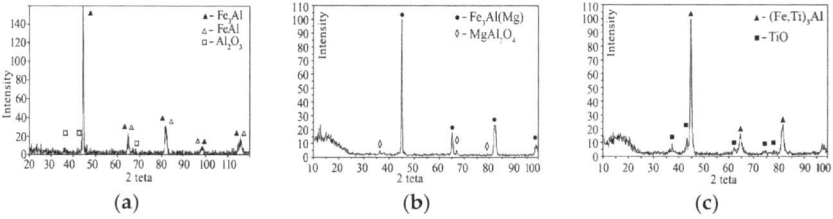

Figure 23. X-ray patterns of Fe$_3$Al-based intermetallic plasma coatings sprayed at I = 500 A: (**a**) Fe$_3$Al, (**b**) Fe-AlMg, (**c**) Fe-TiAl.

Changes in the phase composition of the coatings do not occur when the spraying mode changes.

The characteristics of coatings obtained under different modes of plasma spraying are given in Table 9.

Table 9. Characteristics of Fe$_3$Al-based intermetallic plasma sprayed coatings.

Coating	Current, A	Porosity, %	Microhardness, MPa	Phase Composition
Fe$_3$Al	400	5.2 ± 0.8	4250 ± 760	Fe$_3$Al, FeAl, traces Al$_2$O$_3$
	450	4.2 ± 0.6	4570 ± 780	
	500	3.7 ± 0.3	4790 ± 920	
Fe-AlMg	400	3.8 ± 0.3	4660 ± 560	Solid solution Mg in Fe$_3$Al, MgAl$_2$O$_4$
	450	3.5 ± 0.4	4840 ± 710	
	500	2.6 ± 0.4	4980 ± 550	
Fe-TiAl	400	7.6 ± 1.1	7490 ± 1340	(Fe, Ti)$_3$Al, TiO
	450	5.9 ± 0.8	7860 ± 1200	
	500	4.5 ± 1.0	7980 ± 930	

As expected, the porosity of the coatings decreases (on average by 2%) with an increase in current from 400 to 500 A (Figure 24a), which is associated with a higher kinetic energy of particles upon impact with the substrate and a greater degree of their deformation with increasing current.

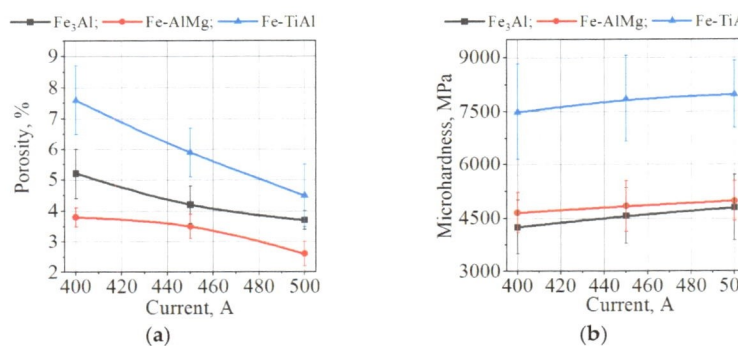

Figure 24. Effect of current change on (**a**) porosity and (**b**) microhardness of Fe$_3$Al-based plasma coatings.

The results of measuring the microhardness of Fe$_3$Al and Fe-AlMg coatings show a slight increase in microhardness relative to the initial MCS powders. This is most likely due to the formation of oxides in the coatings during spraying. In the case of sputtering of the Fe-TiAl powder, an increase in the microhardness of the coating relative to the microhardness of the initial MCS powder by ~2.3 times is noted. This is due to the formation of an intermetallic phase (Fe, Ti)$_3$Al in the coating with additional hardening of the coating by the TiO oxide phase.

From Figure 24b, the increasing current slightly increases the hardness values for all types of coatings related to the content of pores in the coatings. Increasing currents decrease the coatings' porosity, which lead to an increase in microhardness values.

Thus, to obtain plasma coatings with a dense lamellar structure and high adhesion and cohesion strength from Fe$_3$Al-based iron aluminide powders obtained by the MCS method, it is advisable to carry out the spraying process at the following spraying parameters: voltage of 40 V, current of 500 A, plasma–Ar/N$_2$ gas mixture in the ratio of 7.3:1.

4. Conclusions

1. The behaviour of particles of iron aluminides powders (Fe$_3$Al, Fe-AlMg, and Fe-TiAl) obtained by mechanochemical synthesis was analysed at the stage of their transfer through the volume of the high-temperature plasma jet and after impact on the

sprayed surface, depending on the intensity of the arc current in the plasma spraying process;
2. It was established that, as a result of the contact of molten particles with oxygen-containing zones of the plasma jet, aluminium oxide films are formed on the surface of the particles, which acquire a dome shape when the particles are carried by the turbulent plasma jet;
3. It was noted that, during the transfer of particles through the plasma jet, the initial particles are coagulated, resulting in their average size increasing from 9–15 μm to 19–31 μm;
4. An increase in current leads to a 5–10% increase in the degree of particle deformation upon collision with the substrate as a result of the increase in temperature and plasma jet velocity;
5. It has been found that an increase in current from 400 to 500 A during the plasma spraying of powders based on the Fe_3Al intermetallic compound leads to the formation of coatings with a denser structure due to an increase in the degree of particle deformation during the formation of the coating layer;
6. As a result of the research carried out, it was determined that, for the formation of coatings with a thin-lamellar dense structure and a dense boundary with a steel substrate from MCS intermetallic Fe_3Al-based powders, it is advisable to use the following plasma spraying parameters: voltage 40 V, current 500 A, plasma-forming gas–Ar/N_2 mixture at a ratio of 7.3:1.

Author Contributions: Conceptualization, N.V. and C.S.; methodology, N.V., C.S., O.B., O.G. and A.M.; formal analysis, N.V. and O.B.; investigation, N.V., C.S., O.B., O.G., S.S. and A.M.; writing–original draft preparation, N.V., O.B. and C.S.; writing–review and editing, C.S. and N.V.; supervision, C.S. All authors have read and agreed to the published version of the manuscript.

Funding: Financial support from The National Academy of Sciences of Ukraine, Ukraine, Research Project No. 1.6.4.73.21.5, is gratefully acknowledged.

Institutional Review Board Statement: Not applicable.

Informed Consent Statement: Not applicable.

Data Availability Statement: The data are available in a publicly accessible repository.

Acknowledgments: The authors thank the late Yu. Borisov of the E.O. Paton Electric Welding Institute of the National Academy of Sciences of Ukraine for his scientific discussion of the experiments.

Conflicts of Interest: The authors declare no conflict of interest.

References

1. Balasubramaniam, R. Hydrogen in iron aluminides. *J. Alloys Compd.* **2002**, *330*, 506–510. [CrossRef]
2. Cinca, N.; Lima, C.R.C.; Guilemany, J.M. An overview of intermetallics research and application: Status of thermal spray coatings. *J. Mater. Res. Technol.* **2013**, *2*, 75–86. [CrossRef]
3. Moszner, F.; Peng, J.; Suutala, J.; Jasnau, U.; Damani, M.; Palm, M. Application of iron aluminides in the combustion chamber of large bore 2-stroke marine engines. *Metals* **2019**, *9*, 847. [CrossRef]
4. Cinca, N.; Cygan, S.; Senderowski, C.; Jaworska, L.; Dosta, S.; Cano, I.G.; Guilemany, J.M. Sliding wear behaviour of Fe-Al coatings at high temperatures. *Coatings* **2018**, *8*, 268. [CrossRef]
5. Senderowski, C.; Cinca, N.; Dosta, S.; Cano, I.G.; Guilemany, J.M. The effect of hot treatment on composition and microstructure of HVOF iron aluminide coatings in Na_2SO_4 molten salts. *J. Therm. Spray Technol.* **2019**, *28*, 1492–1510. [CrossRef]
6. Song, B.; Dong, S.; Fenineche, N.-E.; Aubry, E.; Grosdidier, T.; Liao, H.; Coddet, C. Microstructure and magnetic properties of atmospheric plasma sprayed Fe–40Al coating obtained from nanostructured powders. *Appl. Phys. A* **2013**, *113*, 787–792. [CrossRef]
7. Yang, D.M.; Tian, B.H. Microstructure and mechanical properties of FeAl coating deposited by low pressure plasma spray. *Appl. Mech. Mater.* **2013**, *333*, 1916–1920. [CrossRef]
8. Mušálek, R.; Kovářík, O.; Skiba, T.; Haušild, P.; Karlík, M.; Colmenares-Angulo, J. Fatigue properties of Fe–Al intermetallic coatings prepared by plasma spraying. *Intermetallics* **2010**, *18*, 1415–1418. [CrossRef]
9. Panas, A.J.; Senderowski, C.; Fikus, B. Thermophysical properties of multiphase Fe-Al intermetallic-oxide ceramic coatings deposited by gas detonation spraying. *Thermochim. Acta* **2019**, *676*, 164–171. [CrossRef]

10. Fikus, B.; Senderowski, C.; Panas, A.J. Modeling of dynamics and thermal history of Fe40Al intermetallic powder particles under gas detonation spraying using propane–air mixture. *J. Therm. Spray Technol.* **2019**, *28*, 346–359. [CrossRef]
11. Chmielewski, T.; Chmielewski, M.; Piatkowska, A.; Grabias, A.; Skowronska, B.; Siwek, P. Structure Evolution of the Fe-Al Arc-Sprayed Coating Stimulated by Annealing. *Materials* **2021**, *14*, 3210. [CrossRef] [PubMed]
12. Zhang, Y.; Matthews, S.; Munroe, P.; Hyland, M. Effect of particle pre-oxidation on Ni and Ni20Cr splat formation during plasma spraying. *Surf. Coat. Technol.* **2020**, *393*, 125849. [CrossRef]
13. Kudinov, V.; Pekshev, P.; Belashchenko, V.; Solonenko, O.; Saphiullin, V. *Plasma Spraying of Coatings*; Nauka: Moscow, Russia, 1990. (In Russian)
14. Tillmann, W.; Khalil, O.; Baumann, I. Influence of direct splat-affecting parameters on the splat-type distribution, porosity, and density of segmentation cracks in plasma-sprayed YSZ coatings. *J. Therm. Spray Technol.* **2021**, *30*, 1015–1027. [CrossRef]
15. Borisova, A.; Timofeeva, I.; Vasil'kovskaya, M.; Burlachenko, A.; Tsymbalistaya, T. Structural and phase transformations in Fe–Al intermetallic powders during mechanochemical sintering. *Powder Metall. Met. Ceram.* **2015**, *54*, 490–496. [CrossRef]
16. Borisov, Y.; Borisova, A.; Burlachenko, A.; Tsymbalistaya, T.; Senderowski, C. Structure and properties of alloyed powders based on Fe$_3$Al intermetallic for thermal spraying produced using mechanochemical synthesis method. *Paton Weld. J.* **2017**, *9*, 33–39. [CrossRef]
17. Verona, S.P.; da Silva, L.R.R.; Setti, D.; Verona, N.; Paredes, R.; Bolsoni Falcao, R.; Santos, M. Flame spraying of Al/Fe$_3$Al-Fe$_3$AlC$_x$ composites powders obtained by vertical ball milling. *Surf. Coat. Technol.* **2022**, *436*, 128276. [CrossRef]
18. Palm, M. Concepts derived from phase diagram studies for the strengthening of Fe–Al-based alloys. *Intermetallics* **2005**, *13*, 1286–1295. [CrossRef]
19. Borysov, Y.; Borysova, A.; Burlachenko, O.; Tsymbalista, T.; Vasylkivska, M.; Byba, E. Composite powders based on FeMoNiCrB amorphizing alloy with additives of refractory compounds for thermal spraying of coatings. *Paton Weld. J.* **2021**, *11*, 38–47. [CrossRef]
20. Borysov, Y.; Krivtsun, I.; Muzhichenko, A.; Lugscheider, E.; Eritt, U. Computer modelling of the plasma spraying process. *Paton Weld. J.* **2000**, *12*, 42–51.
21. Fauchais, P.; Heberlein, J.; Boulos, M. *Thermal Spray Fundamentals. From Powder to Part*; Springer: New York, NY, USA, 2014. [CrossRef]
22. Kulik, A.; Borisov, Y.; Minuchin, A.; Nikitin, M. *Thermal Spraying of Composite Powders*; Mashinostroenie: Saint Petersburg, Russia, 1985. (In Russian)
23. Yuschchenko, K.; Borisov, Y.; Kuznetsov, V.; Korzh, V. *Surface Engineering*; Naukova Dumka: Kyiv, Ukraine, 2007. (In Ukrainian)

Disclaimer/Publisher's Note: The statements, opinions and data contained in all publications are solely those of the individual author(s) and contributor(s) and not of MDPI and/or the editor(s). MDPI and/or the editor(s) disclaim responsibility for any injury to people or property resulting from any ideas, methods, instructions or products referred to in the content.

Article

The Effect of Fe/Al Ratio and Substrate Hardness on Microstructure and Deposition Behavior of Cold-Sprayed Fe/Al Coatings

You Wang [1], Nan Deng [1], Zhenfeng Tong [2] and Zhangjian Zhou [1,*]

[1] School of Materials Science and Engineering, University of Science and Technology Beijing, Beijing 100083, China
[2] School of Nuclear Science and Engineering, North China Electric Power University, Beijing 102206, China
* Correspondence: zhouzhj@mater.ustb.edu.cn

Abstract: Fe/Al composite coatings with compositions of Fe-25 wt.% Al, Fe-50 wt.% Al and Fe-75 wt.% Al were deposited on pure Al and P91 steel plates by a cold spray, respectively. The microstructure of the cross-section of the fabricated coatings was characterized by SEM and EDX. The bonding strength between the coatings and substrates was measured and analyzed. The effects of the Fe/Al ratios and substrate hardness on the deposition behavior were investigated. It was interesting to find fragmented zones in all fabricated coatings, which were composed of large integrated Al particles and small fragmented Al particles. Meanwhile, the fraction of fragmented zones varied with the fraction of the actual Fe/Al ratio. An Fe/Al ratio of 50/50 appeared to be an optimized ratio for the higher bonding strength of coatings. The in situ hammer effect caused by larger and harder Fe particles played an important role in the cold spray process. The substrate with the higher hardness strengthened the in situ hammer effect and further improved the bonding strength.

Keywords: cold spray; Fe/Al composite coatings; Fe/Al ratio; in situ hammer effect; substrate hardness

1. Introduction

Fe/Al coatings are promising candidates for applications in extreme environments due to their excellent properties such as high-temperature strength, corrosion resistance, wear resistance and water vapor resistance [1–6]. Fe/Al coatings have been investigated for the surface protection of aerospace parts, as a tritium permeation barrier in fusion reactors [7], the surface protection of tubes in power plants and the surface protection of microelectronic elements [8]. Fe/Al coatings can be fabricated by different technologies, including spraying [9], hot dipping [10] and electron beaming [11]. Among them, a thermal spray is an efficient and economic method. However, the high operation temperature during thermal spraying may induce problems such as impurities (oxides and unknown iron aluminides), high porosities, thermal residual stress and even microcracks [12–15]. Brittle iron aluminides also easily form during high-temperature processes. These defects significantly deteriorate the properties of the thermal spray coatings.

In order to solve the problems caused by a high temperature during spraying, solid-state deposition technology—namely, a cold spray—has been developed in recent years, which can operate at temperatures much lower than the melting point of the sprayed materials [16]. It utilizes high-pressure compressed gas to propel microsized particles onto a substrate under atmospheric conditions. In this way, incident particles obtain high-velocity impact energy; the kinetic energy is then transferred into plastic deformation energy and thermal energy [17]. In most conditions, the effect of localized thermal softening is greater than working hardening, which leads to adiabatic shear instability [18]. Thus, by relying on the plastic deformation and adiabatic shear instability of both incident particles and substrates, powders can be deposited on and bonded with substrates by cold spraying.

Cold spraying has been successfully applied to fabricating aerospace spars, stringers and frameworks as well as repairing damaged structural components [19–22].

In previous studies, powders with good plastic deformation abilities such as Al, Zn, Cu and Ni were commonly applied to cold spraying [23,24]. With the development of this technology, composite feedstock powders have been introduced into cold spraying in recent decades [13,25–30]. It was noticed that the composite feedstock powders were always composed of soft metals such as Al/Cu, Ti/Al, Ni/Cu and Al/Mg, among others [26,30,31]. Thus, it is interesting to investigate the cold spray deposition behavior of feedstock powders mixed with both soft and hard metals, which may show better mechanical properties and a more promising application prospect.

Compared with soft metal powders (such as Al), Fe powders are difficult to deposit by cold spraying because of the high elastic modulus. The addition of Al powders into the feedstock powders may improve the deposition efficiency of Fe powders. Previous studies about Fe/Al cold-sprayed coatings mostly used an Fe/Al alloy as a feedstock powder [32]. However, dual-phase Fe/Al composite powders are rarely applied in cold spraying as feedstock powders. Wang et al. [9,33] used Fe/Al composite powders as feedstock powders to investigate the coating characteristics and phase transformations during heat treatments. However, the effect of different compositions of Fe/Al coatings have not been widely discussed. Substrates with different hardnesses may influence the effect of cold spraying [17,18]. In addition, Fe/Al coatings are usually sprayed by helium and nitrogen to improve the deposition efficiency [9,33,34], but they are expensive. It is interesting to understand whether Fe/Al coatings can be cold-sprayed using the atmosphere as a propellent gas.

In this work, the atmosphere was used as a cold-sprayed propellent gas in the process of depositing Fe/Al composite coatings onto pure Al and P91 steel substrates, respectively. The effects of the Fe/Al ratio and substrate hardness on the deposition behavior and microstructure of the coatings were investigated. The bonding strength was measured and the bonding mechanism was discussed.

2. Experimental Procedure

2.1. Cold Spraying

Commercial Fe (99.99 wt.% purity, -200 mesh) and Al (99.99 wt.% purity, -325 mesh) powders were used as the feedstock powders. Most of the Fe and Al powders were spherical, as shown in Figure 1.

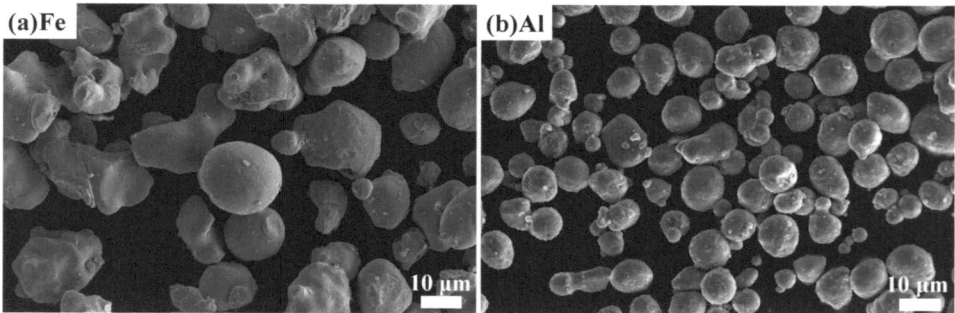

Figure 1. SEM images of feedstock powders: (**a**) Fe and (**b**) Al powders.

Different ratios of Al powders—25 wt.%, 50 wt.% and 75 wt.%—were mechanically mixed with Fe powders in a V-type mixer for 15 h. The mixed Fe/Al powders were preheated by a propelling gas at 350 °C and deposited by a TECHNY LP-TCY-III supersonic cold spraying system onto the substrates. The converging–diverging nozzle of the gun had a throat diameter of 6 mm. The atmosphere gas worked as the propelling and powder-feeding gas. Commercial P91 steel and pure Al plates were used as the substrates. The composition of the commercial P91 steel was Fe-9.00Cr-1.00Mo-0.30Mn-0.20V-0.08C (wt.%) with a hardness of 180 HV; the hardness of the commercial pure Al plates was 85 HV. The constituents of all samples used in this study are listed in Table 1.

Table 1. Constituents of samples (wt.%).

Samples	Fraction of Fe	Fraction of Al	Substrate
Fe-25Al-Al	75	25	
Fe-50Al-Al	50	50	Al
Fe-75Al-Al	25	75	
Fe-25Al-P91	75	25	
Fe-50Al-P91	50	50	P91
Fe-75Al-P91	25	75	

2.2. Shear Bonding Strength Test

The bonding strength was measured by a specially designed shear bonding test, as shown in Figure 2. First, a bar with a half-coating and a half-substrate was cut from a cold-sprayed sample with a diameter of 6 mm and a length of 15 mm, as shown in Figure 2a. Careful wire cutting was then performed from the surface of coating to the coating/substrate interface to form a gap, as well as from the surface of the substrate to the coating/substrate interface to form another gap, as shown in Figure 2b. The distance between these two gaps was 1 mm, as marked by a red line in Figure 2c, which is the front view of Figure 2b. In this way, the whole bar could be considered to be composed of two parts, A and B, connected by the red line marked in Figure 2c. Finally, A and B were separated by applying two forces at both ends of the bar by a CSS-WAW electro-hydraulic servo universal testing machine. The force F was used to calculate the shear bonding strength according to Equation (1):

$$P = F/S \qquad (1)$$

P is the shear bonding strength, MPa; F is the force, N; and S is the contact area, m². In this test, $S = 1 \times 10^{-3} \times 6 \times 10^{-3} = 6 \times 10^{-6}$ m². The shear bonding strength tests were repeated three times for each specimen to ensure the accuracy of results.

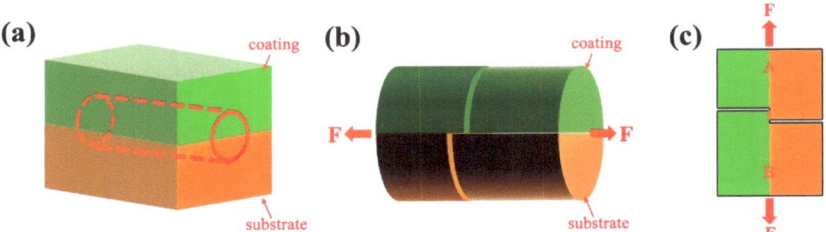

Figure 2. Scheme of shear bonding test, in which the green area represents the Fe/Al coating and the orange area represents the substrate; the bar in (**b**) was cut from (**a**) and the rectangle in (**c**) is the front view of (**b**).

2.3. Characterization

The as-sprayed coated samples were metallographically polished. The microstructure was investigated using field-emission scanning electron microscopy (SEM) (Zeiss Gemini 300 Ultra, Berlin, Germany) equipped with an X-ray spectrometer (EDS) (Oxford Xplore 30 energy-dispersive, Oxford, UK). The fracture surface morphology after the shear bonding strength test was investigated by SEM. An EDS elemental mapping analysis and a line scan analysis were used to measure the compositions. The actual area fractions of the Fe and Al particles in the coatings were estimated by ImageJ Software (Version 1.53, National Institutes of Health, Bethesda, MD, USA) using more than 10 BSE images, which covered the whole cross-section of the specimens. The average area fractions were roughly equal to the volume fractions (vt.%) of the Fe and Al particles in the coatings. They were finally transformed into weight fractions (wt.%) according to m = ρV.

3. Results and Discussion

3.1. Microstructure of the Fe/Al Coating

Figures 3 and 4 show the cross-section morphologies of the Fe/Al coatings with different compositions. In all images, the white particles represent Fe and the grey particles represent Al. These two types of particles were homogenously distributed and showed obvious deformations, as seen in Figure 3a–c. It was also interesting to find that a few dark areas existed, typically along the inter-particle interface, with a different morphology from the porosities or inter-particle boundaries, as indicated by the red box in Figure 3a–c. Two different morphologies, flat zones and rough zones, that contained a number of small fragmented particles less than 100 nm could be observed in these dark areas at higher magnifications, as shown in Figures 3d–f and 4a. The fragmented particles were identified as Al particles from the EDS spot analysis, as shown in Figure 3d–f. Furthermore, the EDS elemental mapping analysis on Fe-75Al-Al confirmed that only the Al element appeared in the whole of the dark areas, which is clearly shown in Figure 4b. The EDS line scan analysis shown in Figure 4c further demonstrated that the content of the Al element in the left flat zone was lower than that in the right rough zone. It should be noted that the results of the line scan analysis were qualitative rather than quantitative due to a low count rate as the black area was lower than the surrounding area. This suggested that the dark areas were composed of Al particles and that they performed with two different morphologies, large integrated particles, and small fractured particles.

This type of morphology (named the fragmented zones by the authors) could have been caused by the high strain rate of the Al particles and the in situ hammer effect of the Fe particles. It has been clarified in previous studies that smaller particles are easier to accelerate and obtain a much higher strain rate [28,31,33]. The Al powders used in this work had a smaller particle size; thus, the high strain rate easily broke the outer layer into fractions. The in situ hammer effect of the Fe particles also promoted this process, which has been confirmed in previous studies [35–38]. Moreover, a high strain rate can aggravate the effect of working hardening. When the effect of working hardening is higher than thermal softening, adiabatic shear instability is weakened, which leads to a loose bonding of these broken Al particles. Therefore, Al particles with fractured outer layers in fragmented zones are easy to peel off and retain a flat and concave morphology, as shown in Figure 4a.

A typical fracture morphology is shown in Figure 5a, in which one large particle with a crater at the upper right was adhered to several small particles. According to the EDS elemental mapping analysis in Figure 5b–d, it was deduced that this was a large Fe particle attached to a few small Al particles. Regarding the crater, there was no direct evidence to identify it as either an Al particle or an Fe particle that had peeled off from the large particle and left a crater. This type of fracture morphology could be attributed to loosen bonding caused by a high strain rate, as mentioned above.

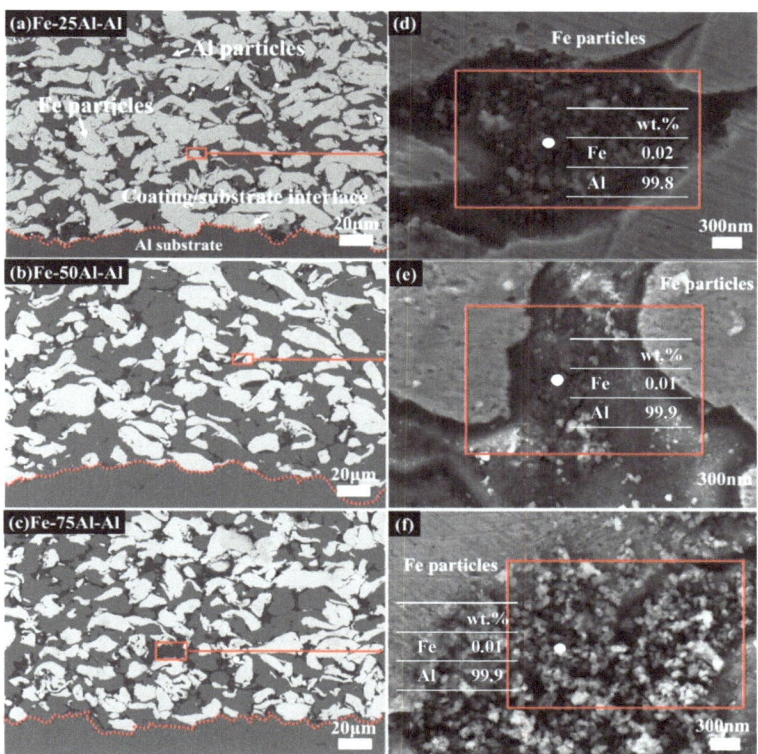

Figure 3. BSE images of cross-section of samples (**a**–**c**) and partial enlarged SEM images (**d**–**f**) with the results of the EDX spot analysis.

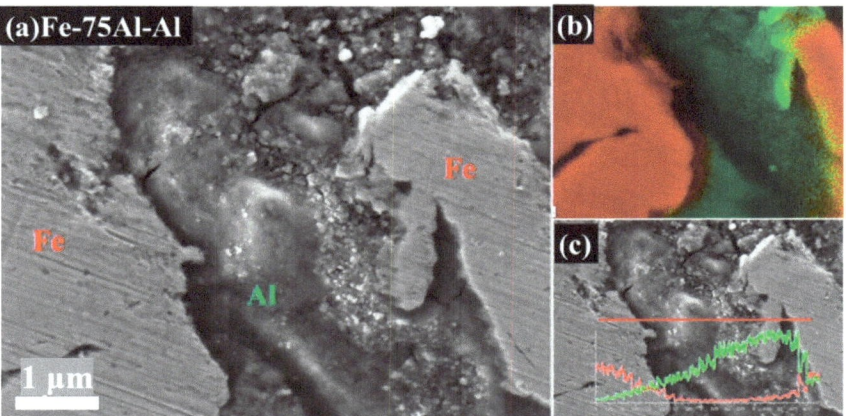

Figure 4. (**a**) SEM image of Fe-75Al-Al; (**b**) EDS mapping elemental analysis (red region is Fe and green region is Al); (**c**) line scan analysis (red and green lines represent the variations in the Fe content and Al content, respectively).

3.2. Effect of the Addition of Al on the Deposition Behavior of the Fe/Al Coating

Pure Fe powders are very difficult to be cold-sprayed onto an Al substrate, especially when using atmosphere as the propellent gas, as reported by previous work [34]. Al powders are one of the most common and efficient materials used in cold spraying due to their high ductility and low elastic modulus [25,30]. The addition of Al powders enabled the dual-phase metal matrix composite powders to form the Fe/Al coating deposited on the Al substrate. The actual fractions of Fe, Al and the fragmented zones were calculated by ImageJ, as mentioned in Section 2.3. Among these, the fraction of the fragmented zones—including large integrated Al particles and small fractured Al particles—was used to indicate the quality of the inter-particle bonding of the cold-sprayed Fe/Al coatings. According to the data listed in Table 2, the actual Fe/Al ratios of Fe-25Al-Al and Fe-50Al-Al were close to the designed compositions whereas the ratio of Fe-75Al-Al was higher than the designed composition, indicating a greater loss of Al compared with the other two during cold spraying. Meanwhile, the fraction of fragmented zones decreased from 10.22% to 7.75%, then rose to 9.62% along with the increase in Al. This trend could be attributed to the in situ hammer effect of the Fe powders [35,36,39]. The density of Fe is much larger than that of Al. The harder and larger Fe powders in the composite feedstock powders hammered the Al powders into small and fractured particles, as mentioned above; thus, the fraction of fragmented zones decreased with the lower actual fraction of Fe powders in Fe-25Al-Al and Fe-50Al-Al.

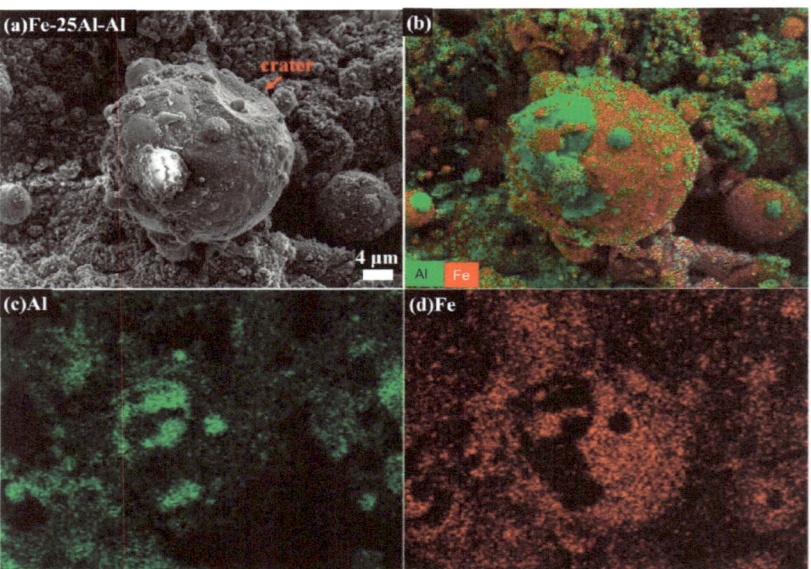

Figure 5. SEM image (**a**) and EDS elemental mapping analysis (**b–d**) of an Fe particle attached to many Al particles in the fracture morphology.

Table 2. Actual fraction of Fe, Al and fragmented zones in Fe/Al coatings (wt.%).

Samples	Actual Fraction			
	Fe	Al	Fragmented Zones	Error
Fe-25Al-Al	60.45	29.33	10.22	±2.01
Fe-50Al-Al	43.28	48.97	7.75	±1.43
Fe-75Al-Al	45.56	44.82	9.62	±1.77
Fe-25Al-P91	66.89	31.95	1.16	±1.25

It was noted that the fraction of Fe was much higher than the designed composition in Fe-75Al-Al and the fraction of fragmented zones increased again when the content of Al increased from 50% to 75%. This result was quite different from expectation, as Fe powders ought to bounce off during cold spraying because of the higher hardness and larger size [40]. However, it might have been a result of the addition of 75 wt.% Al. The Fe powders may have been harder to bounce off when surrounded by sufficient soft Al powders compared with the other two samples. This was in agreement with the experiment of Ng et al. [31]. They found that following a rise in the Al content of samples, Ti6Al4V was tightly surrounded by Al; thus, the ratio of rebound Ti alloy particles decreased. It seemed that the addition of Al effectively prevented the rebound behavior of the Fe powders. Thus, the Fe/Al ratio of Fe-75Al-Al and the fraction of fragmented zones were higher than those of Fe-50Al-Al.

As a result, the effect of the addition of Al powders worked in two ways: the in situ hammer effect and the rebound behavior of the Fe powders. The actual fraction of Fe and fragmented zones had the same trend. In order to ensure a successful deposition and to reduce the loss of feedstock powders, an optimized ratio of Fe/Al during cold spraying on a pure Al substrate is around 50/50.

3.3. Effect of the Substrate Hardness on the Microstructure

Figure 6 shows macro pictures of the as-sprayed samples with different substrates after wire cutting and polishing, in which the coating thickness of Fe-75Al-Al was obviously much lower than the other two samples. This was probably related to its special deposition behavior, as discussed before. It should be noted that the top surface of these coatings was non-uniform because our priority was to spray the coatings as thickly as possible during fabrication rather than obtain a flat surface. In general, it is possible to obtain a rather high thickness at the millimeter level for cold-sprayed Fe/Al coatings. Furthermore, the coatings with a pure Al substrate integrated well whereas the coatings of Fe-50Al-P91 and Fe-75Al-P91 peeled off from the P91 steel substrate during wire cutting. It seemed that only the Fe-rich coating could be successfully deposited onto the P91 steel substrate, indicating that the substrate hardness might have an effect on the deposition behavior.

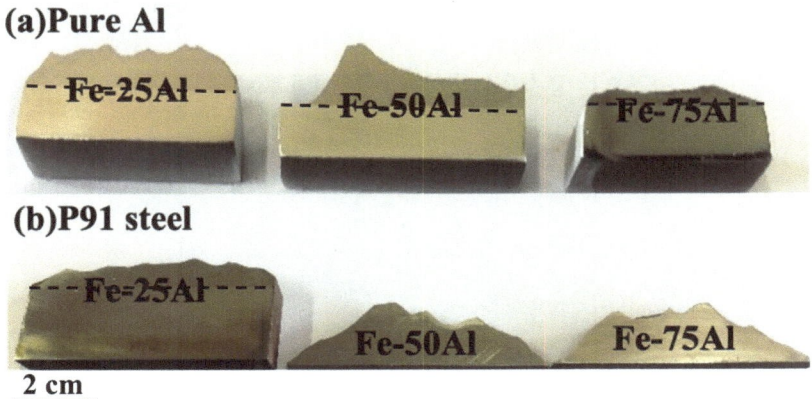

Figure 6. Images of as-sprayed samples after wire cutting and polishing: (**a**) samples with pure Al substrate; (**b**) samples with P91 steel substrate.

The Fe-25Al-Al and Fe-25Al-P91 samples were compared to investigate the effect of the substrate hardness on the deposition behavior, including the microstructure and bonding strength. Figures 7–10 show the SEM images of the Fe-25Al-Al and Fe-25Al-P91 samples. The Fe/Al coatings were dense and the Fe and Al particles were homogeneously distributed, as shown in Figure 7. The fraction of fragmented zones in Fe-25Al-P91 appeared to be much smaller than that in Fe-25Al-Al. The fraction of Al fragmented zones in Fe-25Al-P91 was also calculated by ImageJ Software, as mentioned in Section 2.3, and is listed in Table 2. The value was only 1.16%, which was much lower than the 10.22% of Fe-25Al-Al. It was also reflected by the morphology in Figure 7, in which less fragmented Al particle zones can be seen in Figure 7b than in Figure 7a. The fracture morphology in Figure 8 shows this feature more directly; the light particles are Fe and the dark particles are Al. It was apparent that the Fe particles in Fe-25Al-P91 deformed much more than the other sample. In previous studies, many researchers believed that the hardness of the substrate could only constrain a layer that was several micrometers thick [16,17,30]. Wang [41] also proved that a harder substrate was of benefit when strengthening the inter-particle bonding in an Al_2O_3 coating. The decrease in fragmented zones in Fe-25Al-P91 and a greater number of deformed Fe particles appeared to be further evidence to support this conclusion. Fragmented zones appeared along not only the inter-particle interface but also the coating/substrate interface of Fe-25Al-Al and a narrow gap existed along the coating/substrate interface of Fe-25Al-P91, as shown in Figure 7. Further analyses of the coating/substrate interfaces of the Fe-25Al-Al and Fe-25Al-P91 samples are shown in Figures 9 and 10, respectively.

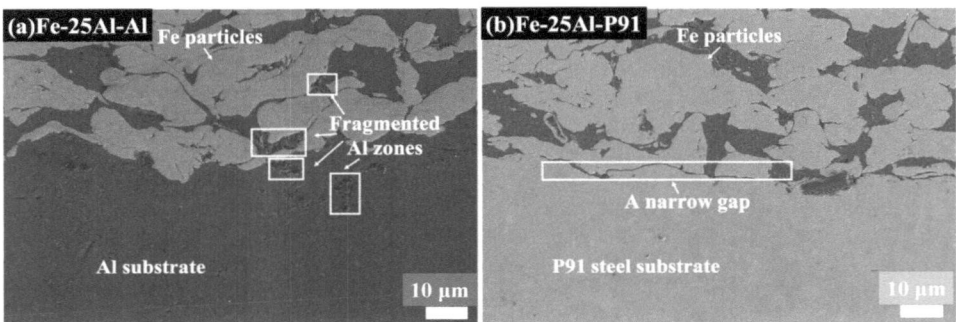

Figure 7. SEM images of Fe-25Al-Al and Fe-25Al-P91, in which fragmented zones appeared along the inter-particle interface and the coating/substrate interface in (**a**). A narrow gap existed along the coating/substrate interface in (**b**).

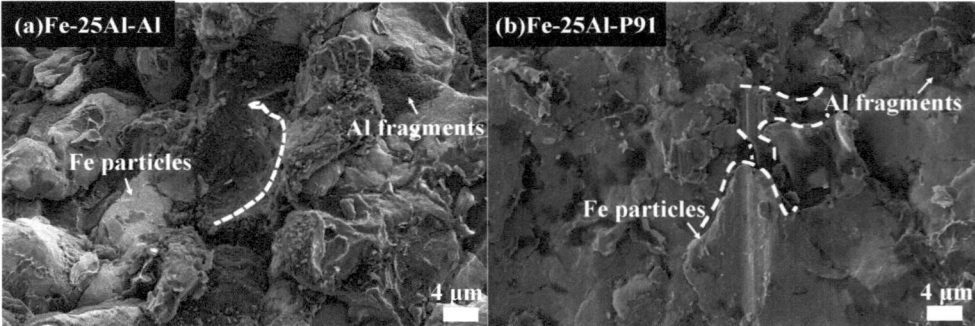

Figure 8. SEM images of the fracture morphology along the inter-particle interface of Fe-25Al-Al and Fe-25Al-P91.

It was noticed that most of the small and fractured Al particles had peeled off and left integrated particles in the fragmented zones along the coating/substrate interface, as shown in Figure 9a. A few microcracks originated and expanded only in these zones, as indicated by the red arrows. The SEM images and EDS elemental mapping analysis results shown in Figure 9b–d proved that the composition of the fragmented zones along the interface was also Al elements. It appeared that these fragmented zones along the interface in Fe-25Al-Al were not continuous. In contrast, the narrow gap, with a less than 1 μm width, in Fe-25Al-P91 was continuous along the interface, as shown in Figure 10b. The EDS line scan analysis showed both small fractured Fe and Al particles existed in the gap, as shown in Figure 10a. We believed that the formation of these two different morphologies along the interface was induced by the in situ hammer effect. The authors believed that the in situ hammer effect could be divided into two parts: hammering and densifying. Regarding Fe-25Al-Al, the hammering effect was strong due to the high content of Fe powders, but the densifying effect was relative weak because of the soft Al substrate. Thus, microcracks originated in the fragmented zones along the coating/substrate interface, which might have been caused by the accumulated stress. Regarding Fe-25Al-P91, the cooperated effect of hammering and densifying promoted the formation of the gap consisting of Fe and Al particles. During the process of cold spraying, high-speed Al powders reached the interface first. The Fe powders then crushed the Al particles into nanosized fragments by hammering. Finally, cooperating with the hard P91 steel substrate, the subsequent Fe powders densified the fragmented Al particles into a narrow gap at the interface, which worked as a transition layer to further improve the bonding strength. The densifying effect was weakened in the samples with fewer Fe particles (such as Fe-50Al-P91 and Fe-75Al-P91) with peeling-off coatings. In addition, due to the high hardness of both the Fe powders and P91 substrates, the high-speed Fe powders were easier to bounce off whilst being sprayed onto the P91 substrate than that onto the Al substrate. The effect of in situ hammering was weakened and led to the peeling off of the coatings on Fe-50Al-P91 and Fe-75Al-P91, as shown in Figure 6. In this way, it could be deduced that the substrate hardness had an impact on the deposition behavior.

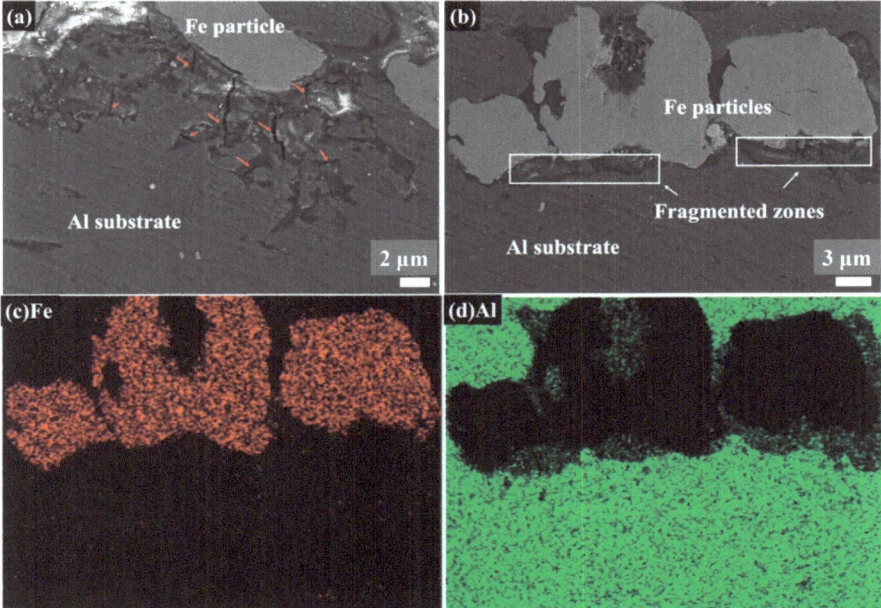

Figure 9. SEM images and EDS elemental mapping along the coating/substrate interface of Fe-25Al-Al; microcracks originated and expanded in the fragmented zones in (**a**) whereas non-continuous fragmented zones formed along the interface in (**b**–**d**).

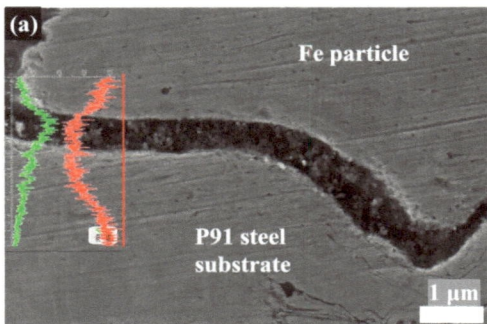
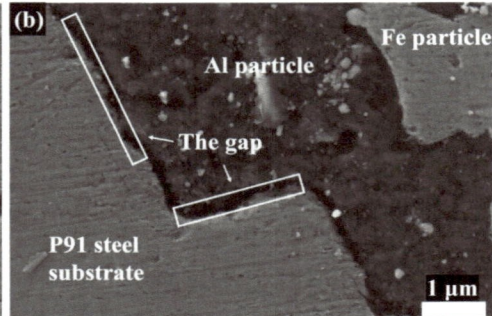

Figure 10. SEM images and EDS line scan analysis of coating/substrate interface of Fe-25Al-P91. (**a**) The green line is the trend curve of Al content and the red line is the trend curve of Fe content. (**b**) shows the enlarged images of the gap.

3.4. Effect of the Fe/Al Ratio and Substrate Hardness on the Bonding Strength

The results of the bonding strength tests are listed in Table 3, in which the value of the bonding strength was the average value of three repeated tests. The variation trend of the bonding strength was the same as the trend of the actual Al fraction in Fe-25Al-Al, Fe-50Al-Al and Fe-75Al-Al. It may have been a result of the in situ hammer effect on the coating/substrate interface, including hammering and densifying, as mentioned in Section 3.3. The addition of Al significantly decreased the hammering effect and thus decreased defects such as microcracks along the coating/substrate interface, as shown in Figure 9a. Therefore, the bonding strength was improved. Regarding the Fe-25Al-P91 sample, its bonding strength of 109 MPa was higher than that of Fe-25Al-Al. However, with an increasing content of Al, the densifying effect was weakened and the bonding strengths of the coatings of Fe-50Al-Al and Fe-75Al-Al decreased. The fracture morphology along the coating/substrate interface after the shear bonding strength test is shown in Figure 11. Traces of the dropped Fe particles on the interface can be observed, marked by white dashed lines in Figure 11a, whereas the fragmented Al particles remained after the shearing testing, as shown in Figure 11b. This type of fracture morphology might be attributed to the higher hardness of the P91 steel substrate as the higher hardness of the substrate strengthened the densifying effect induced by the in situ hammer effect. During the process, the narrow gap at the interface of Fe-25Al-P91, as shown in Figure 10b, acted as the transition layer to further improve the bonding strength.

Table 3. A summary of the actual fraction and bonding strength of all samples.

Samples	Actual Fraction (wt.%)			Bonding Strength (MPa)
	Fe	Al	Error	
Fe-25Al-Al	60.45	39.55	±2.01	52
Fe-50Al-Al	43.28	56.72	±1.43	73
Fe-75Al-Al	45.56	54.44	±1.77	65
Fe-25Al-P91	66.89	33.11	±2.13	109

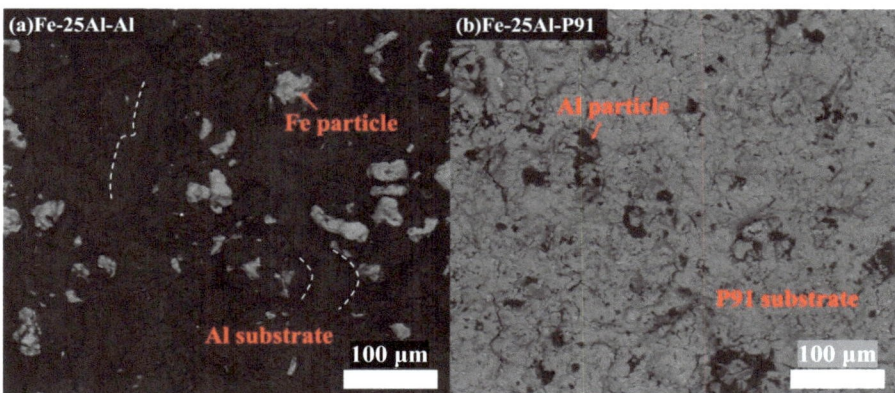

Figure 11. SEM images of fracture morphology along the coating/substrate interface of Fe-25Al-Al and Fe-25Al-P91 after shear bonding tests.

4. Conclusions

Fe/Al coatings with different composition designs were fabricated onto pure Al and P91 steel substrates by cold spraying. The effects of the Fe/Al ratio and substrate hardness on the deposition behavior and bonding strength were investigated. The main conclusions can be summarized as follows:

1. Samples with three different Fe/Al ratios all showed a special morphology of fragmented zones with Al elements. The fragmented zones were composed of large integrated Al particles and small fractured Al particles.
2. The Fe/Al ratio showed a significant influence on the deposition behavior of the cold-sprayed coatings. Fe/Al coatings with different Fe/Al ratios could be successfully deposited onto an Al substrate by cold spraying; a coating with an Fe/Al ratio of 50/50 (wt.%) showed a relatively high bonding strength. For the P91 substrate, only a coating with an Fe/Al ratio of 75/25 (wt.%) could be successfully deposited.
3. The hardness of the substrate had an obvious influence on both the deposition behavior and the bonding strength. Influenced by the in situ hammer effect, the fragmented zones were densified to a transition layer and further improved the bonding strength of Fe-25Al-P91.

Author Contributions: Conceptualization, Y.W. and Z.Z.; methodology, N.D.; investigation, Y.W.; writing—original draft preparation, Y.W.; writing—review and editing, Z.T. and Z.Z.; supervision, Z.Z.; project administration, Z.Z.; funding acquisition, Z.T. and Z.Z. All authors have read and agreed to the published version of the manuscript.

Funding: The authors would like to thank the National Natural Science Foundation of China (No. U1967212) and the National Magnetic Confinement Fusion Energy Research Project of China (No. 2018YFE030610) for their financial support.

Institutional Review Board Statement: Not applicable.

Informed Consent Statement: Not applicable.

Data Availability Statement: The data that support the findings of this study are available on request from the corresponding author, Z. Z., upon reasonable request.

Acknowledgments: The authors would like to thank Qingya Li from Shiyanjia Lab (www.shiyanjia.com (accessed on: 1 March 2021)) for the SEM analysis.

Conflicts of Interest: The authors declare no conflict of interest.

References

1. Zeina, E.C.; Kassem, W.; Shehadeh, M.; Hamade, R.F. On the Mechanical Response and Intermetallic Compound Formation in Al/Fe Interface: Molecular Dynamics Analyses. *Philos. Mag.* **2020**, 1–20. [CrossRef]
2. Wang, Y.; Deng, N.; Zhou, Z.J.; Tong, Z.F. Influence of heating temperature and holding time on the formation sequence of iron aluminides at the interface of Fe:Al coatings. *Mater. Today Commun.* **2021**, *28*, 102516. [CrossRef]
3. Janovska, M.P.; Sedlak, J.C.; Koller, M.; Siska, F.; Seiner, F. Characterization of Bonding Quality of a Cold-Sprayed Deposit by Laser Resonant Ultrasound Spectroscopy. *Ultrasonics* **2020**, *106*, 106140. [CrossRef]
4. Wang, H.R.; Tyler, H.; Zhu, C.Y.; Kenneth, S.V. Design, Fabrication and Characterization of Feal-Based Metallic-Intermetallic Laminate (Mil) Composites. *Acta. Mater.* **2019**, *175*, 445–456. [CrossRef]
5. Peska, M.; Karczewski, K.; Rzeszotarska, M.; Polanski, M. Direct Synthesis of Fe-Al Alloys from Elemental Powders Using Laser Engineered Net Shaping. *Materials* **2020**, *13*, 531. [CrossRef]
6. Karimi, H.; Morteza, H. Effect of Sintering Techniques on the Structure and Dry Sliding Wear Behavior of Wc-Feal Composite. *Ceram. Int.* **2020**, *11*, 18487. [CrossRef]
7. Hu, L.; Zhang, G.; Wang, H.; Yang, F.; Xiang, X.; Hu, M.; Cao, J.; Tang, T. Optimum preparation of Fe-Al/α-Al$_2$O$_3$ coating on 21-6-9 austenitic stainless steel. *Fusion Eng. Des.* **2019**, *148*, 111280. [CrossRef]
8. Haidara, F.; Record, M.-C.; Duployer, B.; Mangelinck, D. Phase formation in Al–Fe thin film systems. *Intermetallics* **2012**, *23*, 143–147. [CrossRef]
9. Wang, H.T.; Ji, G.C.; Li, C.J. Characterization of microstructure and properties of nanostructured Fe-Al/WC intermetallic composite coating deposited by cold spraying. In Proceedings of the 2010 International Conference on Mechanic Automation and Control Engineering (MACE), Wuhan, China, 26–28 June 2010; pp. 4–7.
10. Cheng, W.-J.; Wang, C.-J. Growth of intermetallic layer in the aluminide mild steel during hot-dipping. *Surf. Coat. Technol.* **2009**, *204*, 824–828. [CrossRef]
11. Wei, D.; Xin, W. Study on the Microstructure Modification of 6061Aluminum by Electron Beam Surface Treatment. Proceedings of 2010 International Conference on Measuring Technology and Mechatronics Automation, Changsha, China, 13–14 March 2010; pp. 665–668.
12. Senderowski, C.; Bojar, Z. Gas detonation spray forming of Fe–Al coatings in the presence of interlayer. *Surf. Coat. Technol.* **2008**, *202*, 3538–3548. [CrossRef]
13. Wong, W.; Irissou, E.; Ryabinin, A.N.; Legoux, J.-G.; Yue, S. Influence of Helium and Nitrogen Gases on the Properties of Cold Gas Dynamic Sprayed Pure Titanium Coatings. *J. Therm. Spray Technol.* **2010**, *20*, 213–226. [CrossRef]
14. Duan, B.; Shen, T.; Wang, D. Effects of solid loading on pore structure and properties of porous FeAl intermetallics by gel casting. *Powder Technol.* **2019**, *344*, 169–176. [CrossRef]
15. Liu, Y.-N.; Sun, Z.; Cai, X.-P.; Jiao, X.-Y.; Feng, P.-Z. Fabrication of porous FeAl-based intermetallics via thermal explosion. *Trans. Nonferrous Met. Soc. China* **2018**, *28*, 1141–1148. [CrossRef]
16. Arabgol, Z.; Villa Vidaller, M.; Assadi, H.; Gärtner, F.; Klassen, T. Influence of thermal properties and temperature of substrate on the quality of cold-sprayed deposits. *Acta Mater.* **2017**, *127*, 287–301. [CrossRef]
17. Yin, S.; Wang, X.-F.; Li, W.Y.; Jie, H.-E. Effect of substrate hardness on the deformation behavior of subsequently incident particles in cold spraying. *Appl. Surf. Sci.* **2011**, *257*, 7560–7565. [CrossRef]
18. Lee, W.-S.; Chen, T.-H.; Lin, C.-F.; Lu, G.-T. Adiabatic Shearing Localisation in High Strain Rate Deformation of Al-Sc Alloy. *Mater. Trans.* **2010**, *51*, 1216–1221. [CrossRef]
19. Xiong, Y.; Zhuang, W.; Zhang, M. Effect of the thickness of cold sprayed aluminium alloy coating on the adhesive bond strength with an aluminium alloy substrate. *Surf. Coat. Technol.* **2015**, *270*, 259–265. [CrossRef]
20. Rech, S.; Trentin, A.; Vezzù, S.; Vedelago, E.; Legoux, J.-G.; Irissou, E. Different Cold Spray Deposition Strategies: Single- and Multi-layers to Repair Aluminium Alloy Components. *J. Therm. Spray Technol.* **2014**, *23*, 1237–1250. [CrossRef]
21. Moridi, A.; Gangaraj, S.M.H.; Vezzu, S.; Guagliano, M. Number of Passes and Thickness Effect on Mechanical Characteristics of Cold Spray Coating. *Procedia Eng.* **2014**, *74*, 449–459. [CrossRef]
22. Dykhuizen, R.C.; Smith, M.F. Gas Dynamic Principles of Cold Spray. *J. Therm. Spray Technol.* **1998**, *7*, 205–212. [CrossRef]
23. Panda, D.; Kumar, L.; Alam, S.N. Development of Al-Fe3Al Nanocomposite by Powder Metallurgy Route. *Mater. Today Proc.* **2015**, *2*, 3565–3574. [CrossRef]
24. Cinca, N.; Drehmann, R.; Dietrich, D.; Gärtner, F.; Klassen, T.; Lampke, T.; Guilemany, J.M. Mechanically induced grain refinement, recovery and recrystallization of cold-sprayed iron aluminide coatings. *Surf. Coat. Technol.* **2019**, *380*, 125069. [CrossRef]
25. Van Steenkiste, T.H.; Smith, J.R.; Teets, R.E. Aluminum coatings via kinetic spray with relatively large powder particles. *Teets Surf. Coat. Technol.* **2002**, *154*, 237–252. [CrossRef]
26. Koivuluoto, H.; Lagerbom, J.; Vuoristo, P. Microstructural Studies of Cold Sprayed Copper, Nickel, and Nickel-30% Copper Coatings. *J. Therm. Spray Technol.* **2007**, *16*, 488–497. [CrossRef]
27. Canales, H.; Cano, I.G.; Dosta, S. Window of deposition description and prediction of deposition efficiency via machine learning techniques in cold spraying. *Surf. Coat. Technol.* **2020**, *401*, 126143. [CrossRef]
28. Li, Y.-J.; Luo, X.-T.; Li, C.-J. Improving deposition efficiency and inter-particle bonding of cold sprayed Cu through removing the surficial oxide scale of the feedstock powder. *Surf. Coat. Technol.* **2021**, *407*, 126709. [CrossRef]
29. Kang, H.-K.; Kang, S.B. Tungsten/copper composite deposits produced by a cold spray. *Scr. Mater.* **2003**, *49*, 1169–1174. [CrossRef]

30. Richer, P.; Jodoin, B.; Ajdelsztajn, L.; Lavernia, E.J. Substrate Roughness and Thickness Effects on Cold Spray Nanocrystalline Al-Mg Coatings. *J. Therm. Spray Technol.* **2006**, *15*, 246–254. [CrossRef]
31. Ng, C.-H.; Yin, S.; Lupoi, R.; Nicholls, J. Mechanical reliability modification of metal matrix composite coatings by adding Al particles via cold spray technology. *Surf. Interfaces* **2020**, *20*, 100515. [CrossRef]
32. Cinca, N.; List, A.; Gärtner, F.; Guilemany, J.M.; Klassen, T. Coating formation, fracture mode and cavitation performance of Fe40Al deposited by cold gas spraying. *Surf. Eng.* **2014**, *31*, 853–859. [CrossRef]
33. Wang, H.-T.; Li, C.-J.; Yang, G.-J.; Li, C.-X. Cold spraying of Fe/Al powder mixture: Coating characteristics and influence of heat treatment on the phase structure. *Appl. Surf. Sci.* **2008**, *255*, 2538–2544. [CrossRef]
34. Chu, X.; Che, H.; Vo, P.; Chakrabarty, R.; Sun, B.; Song, J.; Yue, S. Understanding the cold spray deposition efficiencies of 316L/Fe mixed powders by performing splat tests onto as-polished coatings. *Surf. Coat. Technol.* **2017**, *324*, 353–360. [CrossRef]
35. Wei, Y.-K.; Li, Y.-J.; Zhang, Y.; Luo, X.-T.; Li, C.-J. Corrosion resistant nickel coating with strong adhesion on AZ31B magnesium alloy prepared by an in-situ shot-peening-assisted cold spray. *Corros. Sci.* **2018**, *138*, 105–115. [CrossRef]
36. Luo, X.-T.; Wei, Y.-K.; Wang, Y.; Li, C.-J. Microstructure and mechanical property of Ti and Ti6Al4V prepared by an in-situ shot peening assisted cold spraying. *Mater. Des.* **2015**, *85*, 527–533. [CrossRef]
37. Wei, Y.-K.; Luo, X.-T.; Li, C.-X.; Li, C.-J. Optimization of In-Situ Shot-Peening-Assisted Cold Spraying Parameters for Full Corrosion Protection of Mg Alloy by Fully Dense Al-Based Alloy Coating. *J. Therm. Spray Technol.* **2016**, *26*, 173–183. [CrossRef]
38. Wei, Y.-K.; Luo, X.-T.; Chu, X.; Huang, G.-S.; Li, C.-J. Solid-state additive manufacturing high performance aluminum alloy 6061 enabled by an in-situ micro-forging assisted cold spray. *Mater. Sci. Eng. A* **2020**, *776*, 139024. [CrossRef]
39. Chakradhar, R.P.S.; Chandra Mouli, G.; Barshilia, H.; Srivastava, M. Improved Corrosion Protection of Magnesium Alloys AZ31B and AZ91 by cold-sprayed aluminum coatings. *J. Therm. Spray Technol.* **2021**, *30*, 371–384. [CrossRef]
40. Winnicki, M.; Kozerski, S.; Małachowska, A.; Pawłowski, L.; Rutkowska-Gorczyca, M. Optimization of ceramic content in nickel–alumina composite coatings obtained by low pressure cold spraying. *Surf. Coat. Technol.* **2021**, *405*, 126732. [CrossRef]
41. Wang, L.-S.; Zhou, H.-F.; Zhang, K.-J.; Wang, Y.-Y.; Li, C.-X.; Luo, X.-T.; Yang, G.-J.; Li, C.-J. Effect of the powder particle structure and substrate hardness during vacuum cold spraying of Al_2O_3. *Ceram. Int.* **2017**, *43*, 4390–4398. [CrossRef]

Disclaimer/Publisher's Note: The statements, opinions and data contained in all publications are solely those of the individual author(s) and contributor(s) and not of MDPI and/or the editor(s). MDPI and/or the editor(s) disclaim responsibility for any injury to people or property resulting from any ideas, methods, instructions or products referred to in the content.

Article

Impact of an Aluminization Process on the Microstructure and Texture of Samples of Haynes 282 Nickel Alloy Produced Using the Direct Metal Laser Sintering (DMLS) Technique

Jarosław Mizera [1,*], Bogusława Adamczyk-Cieślak [1], Piotr Maj [1], Paweł Wiśniewski [1], Marcin Drajewicz [2] and Ryszard Sitek [1]

1. Faculty of Materials Science and Engineering, Warsaw University of Technology, Wołoska 141, 02-507 Warsaw, Poland
2. Department of Materials Science, Faculty of Mechanical Engineering and Aeronautics, Rzeszów University of Technology, Al. Powstańców Warszawy 12, 35-959 Rzeszów, Poland
* Correspondence: jaroslaw.mizera@pw.edu.pl

Abstract: In this study, we examined the effects of an aluminization process on the microstructure and texture of Haynes 282 nickel samples fabricated using the direct metal laser sintering technique. The aluminization process involved the use of chemical vapor deposition with $AlCl_3$ vapors in a hydrogen atmosphere at a temperature of 1040 °C for 8 h. Following the 3D printing and aluminization steps, we analyzed the microstructure of the Haynes 282 nickel alloy samples using light microscopy and scanning electron microscopy. Additionally, we investigated the texture using X-ray diffractometry. A texture analysis revealed that after the process of direct laser sintering of metals, the texture of the Haynes 282 nickel alloy samples developed a texture typical of cast materials. Then, in the aluminization process, the texture was transformed—from foundry-type components to a texture characteristic of recrystallization.

Keywords: Haynes 282 nickel alloy; DMLS; texture; CVD; aluminide layer

Citation: Mizera, J.; Adamczyk-Cieślak, B.; Maj, P.; Wiśniewski, P.; Drajewicz, M.; Sitek, R. Impact of an Aluminization Process on the Microstructure and Texture of Samples of Haynes 282 Nickel Alloy Produced Using the Direct Metal Laser Sintering (DMLS) Technique. *Materials* 2023, *16*, 5108. https://doi.org/10.3390/ma16145108

Academic Editor: Antonio Santagata

Received: 1 June 2023
Revised: 7 July 2023
Accepted: 17 July 2023
Published: 20 July 2023

Copyright: © 2023 by the authors. Licensee MDPI, Basel, Switzerland. This article is an open access article distributed under the terms and conditions of the Creative Commons Attribution (CC BY) license (https://creativecommons.org/licenses/by/4.0/).

1. Introduction

Haynes 282 is a nickel alloy that is widely used in the aerospace industry because of its outstanding mechanical properties at high temperatures [1], which are mainly attributed to the strengthening effect of the γ′ phase and solution hardening. Both of these mechanisms significantly decrease the mobility of dislocations at high temperatures [2]. The alloy has a large content of Cr and other refractory metals, which increases its oxidation resistance and chemical stability [3]. Thanks to these advantages and its relatively low price, Haynes 282 is commonly used in demanding applications. Laser powder bed fusion (LPBF) is a modern process that involves sintering metallic powders, layer by layer, using a high-power laser [4]. The process is conducted in a protective atmosphere, which decreases oxidation and contamination of the powder. The main advantages of this method are tool-free machining to the desired shape and size, relatively high strength, and the ability to machine materials that can be problematic using conventional manufacturing methods. The disadvantages are the relatively high costs of the equipment and the powder material, long manufacturing times, and a non-equilibrium structure that is susceptible to cracking and porosity defects [5]. There is a plethora of research concerning an enormous variety of materials produced this way; they include nickel alloys, steels, titanium, composite refractory materials, and others [6–8]. The laser powder bed fusion (LPBF) of nickel-based alloys is a widely used process that results in a characteristic microstructure with dendritic meltpools. Phase segregation is an undesirable feature that decreases a material's wear properties at high temperatures, which is especially unacceptable for uses in aviation [9]. As a remedy, heat treatment is used. The material is supersaturated and aged at a specific temperature for a defined length of time, depending on the alloy. It is worth noting that the primary structure after LPBF significantly affects the properties after heat treatment as well [10]. In numerous research papers, the key mechanical properties are significantly lower than the wrought material. This is most probably a result of defects and heterogeneities imposed in the LPBF [11–13]. For this reason, additional treatments are used to increase the resistance of the material [14–16].

One promising avenue for advancing the structural integrity and surface properties of nickel alloys fabricated through 3D printing techniques is the implementation of high-temperature thermo-chemical treatments. These treatments aim to enhance structural homogeneity while improving surface characteristics. By employing surface engineering techniques to create corrosion-resistant layers at elevated temperatures, the range of applications for nickel alloys manufactured using additive techniques can be significantly expanded, particularly in the demanding high-temperature environments seen in the energy and aviation industries. In this context, the β-NiAl intermetallic phase presents a particularly advantageous set of properties, including a high melting point, low density, and high elasticity modulus [17,18]. Exploiting these properties, the objective of this study was to investigate the effects of a high-temperature, low-activity aluminization process on the microstructure and texture of Haynes 282 nickel alloy samples produced using the direct metal laser sintering (DMLS) technique.

The aim of this study was to investigate microstructural changes resulting from the aluminization process. It is widely recognized that surface structure plays a significant role in chemical vapor deposition (CVD). In this research, our main focus was on analyzing the texture of the resulting layer and its impact on the morphology of both the layer itself and the 3D-printed material obtained through direct metal laser sintering (DMLS). Remarkably, there is a lack of previous research that specifically examines the use of a combination of aerospace materials in such a context.

By examining the microstructural changes induced by the aluminization process, we sought to gain insight into the effects of surface structure on CVD. We investigated both the surface texture of the layer and the characteristics of the DMLS printed material. Our study seeks to bridge the gap in the existing literature by providing a comprehensive analysis of the combined application of aerospace materials in this particular context.

2. Materials and Methods

Sample Preparation and Thermo-Chemical Treatment

Samples 20 × 20 × 20 mm in size were produced from a powder of Haynes 282 nickel alloy, obtained from the company Höganäs for the EOS M 100 printer operating in DMLS technology. The process of producing the sample was carried out using the following parameters: laser power P—80 W; laser velocity V—800 mm/s; distance between successive paths H—0.05 mm; layer thickness D—0.02 mm. After the 3D printing process, the surfaces of the samples were polished with grade 800 sandpaper. Then, 12 h before the aluminization process, the surfaces of the samples were sandblasted to activate them (using particles of α-Al_2O_3 powder) and then cleaned in an ultrasonic washer in ethyl alcohol. On those as-prepared surfaces, we carried out a low-activity process of aluminization using the chemical vapor deposition (CVD) in $AlCl_3$ vapors in an atmosphere of hydrogen as the carrier gas, at a temperature of 1040 °C for 8 h. Furthermore, it is important to highlight that the temperature falls within a range suitable for solution heat treatment. This specific heat treatment process is designed to promote a homogenization of the materials. The metallographic specimens to be investigated were polished with sandpaper of grades 320, 600, and 1200 and then further polished using diamond suspensions of grade 3 μm, followed by 1 μm. The samples for the microstructural observations were etched in Kalings chemical reagent. The observations were made using a Zeiss Axio Scope A1 metallographic microscope and a Hitachi TM-1000 scanning electron microscope. A quantitative analysis of the texture was performed on the basis of four incomplete pole figures (1 1 1), (2 0 0), (2 2 0), and (3 1 1). The measurements were made using a Bruker D8 Discover X-ray diffractometer using filtered CoKα radiation and a spot beam about 1 mm in diameter. Based on the pole figures measures, orientation distribution functions (ODF) were calculated for each sample, and the contributions of the main texture components were quantified.

3. Research Results and Discussion

3.1. Microstructure

Figure 1 shows the microstructure of the Haynes 282 nickel alloy samples in the initial state, as well as after being produced using the EOS M 100 printer operating in DMLS technology. In the initial state (Figure 1a), within the microstructure of the Haynes 282 alloy, there are visible grains of austenite and twins. The microstructure of the sample produced in the Y-Z building plane (Figure 1b) consists of characteristic layers of melt pools. The visible melting pools overlap one another, leaving no empty spaces between them, which suggests that the samples produced have strong mechanical properties. The resulting microstructure is typical of nickel alloys produced using 3D printing.

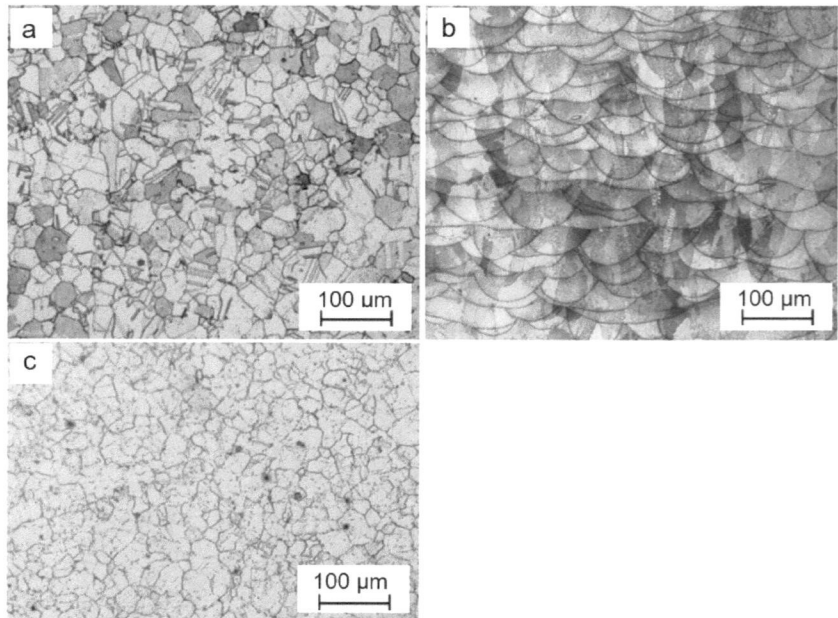

Figure 1. Microstructure of Haynes 282 samples: as received state—(**a**); produced using DMLS in the Y-Z plane—(**b**); annealed at 1040 °C for 8 h—(**c**).

For comparison, a 3D-printed Haynes 282 sample was subjected to a heat treatment process resembling aluminization, at a temperature of 1040 °C for 8 h. Surprisingly, the results exhibit striking similarities, indicating that temperature and diffusion play pivotal roles in shaping the final microstructure. Notably, both cases demonstrated a distinct influence of recrystallization, thereby highlighting its significance in the observed phenomena. The microstructures obtained from these experiments are visually depicted in Figure 1a,c for ease of reference and analysis.

A cross-section of the layer created during the CVD process is shown in Figure 2a. The layer has a complex structure in which two main zones can be distinguished: an exterior β-NiAl zone about 14 μm thick and an interior zone about 15 μm thick, formed of precipitates rich in chromium and other elements of the substrate. The sublayers are divided by a thin zone of Al_2O_3 (residue from the sandblasting process). The arrangement of the layers is typical for the process of aluminizing nickel alloys using the CVD method with AlCl$_3$ vapors [19,20]. The microstructure of the substrate after the aluminization process is presented in Figure 2b. No melting pools characteristic of the DMLS process are visible; the material has undergone recrystallization, and grain boundaries have appeared.

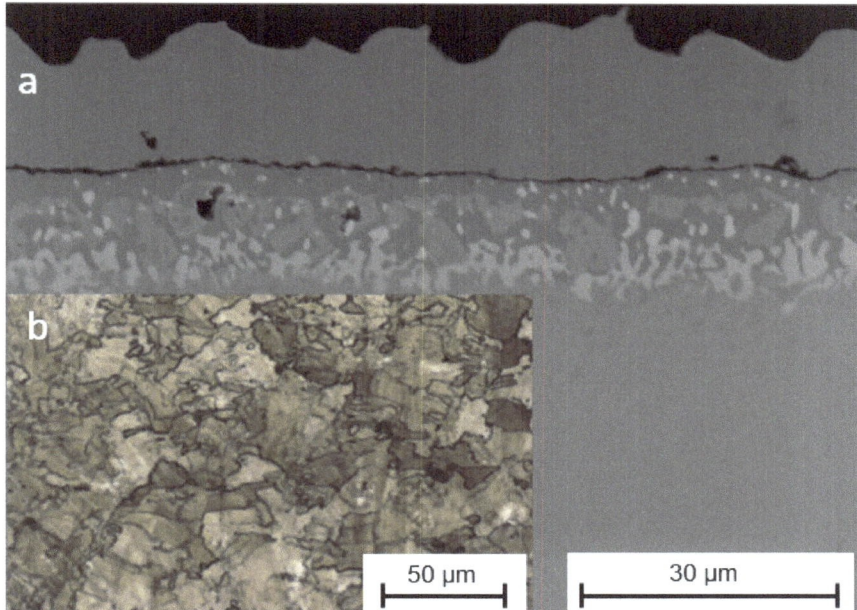

Figure 2. Cross-section of Haynes 282 nickel alloy produced using DMLS and subjected to an aluminization process: microstructure of the layer—(**a**); microstructure of the substrate—(**b**).

3.2. Texture

3.2.1. Texture of Haynes 282 Nickel Alloy Samples in the Initial State (as Received from the Producer)

The experimental, incomplete pole figures (IPF), complete pole figures (CPF), and orientation distribution functions (ODF), along with the identified contributions of particular texture components for the Haynes 282 nickel alloy sample in the initial state, are presented in Figures 3–5 and Table 1. The image of the initial texture revealed in the CPF and ODF, presented in Figure 5, clearly indicates the axial character of the grain orientations identified in the sample after casting and thermal treatment—they are dispersed around the <1 0 0> fiber, which is characteristic of casting states.

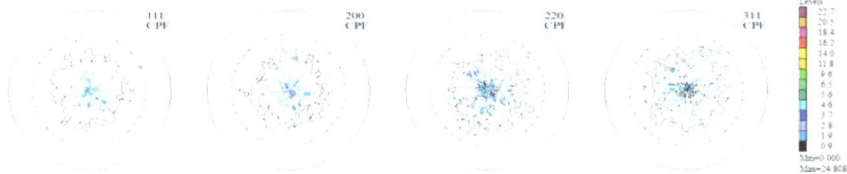

Figure 3. Experimental, incomplete pole figures (1 1 1), (2 0 0), (2 2 0), and (3 1 1) for the sample of Haynes 282 in the initial state.

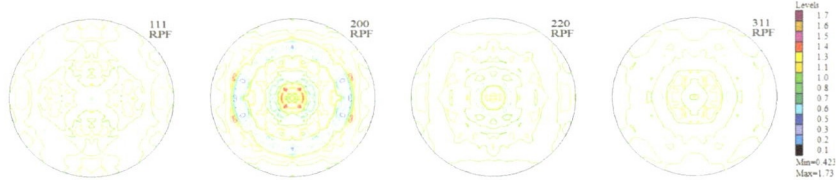

Figure 4. Complete pole figures (1 1 1), (2 0 0), (2 2 0), and (3 1 1) for the sample of Haynes 282 in the initial state.

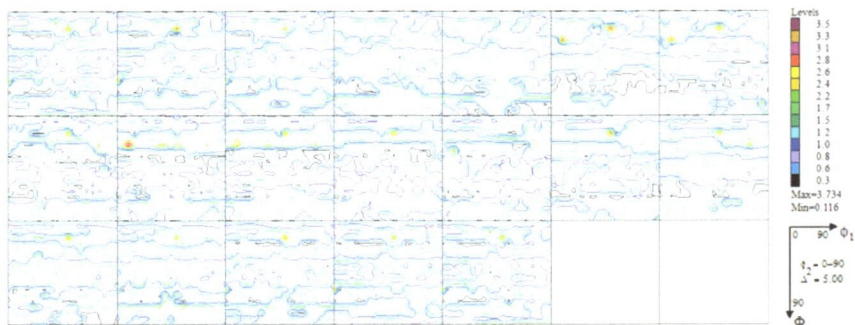

Figure 5. Orientation distribution functions maps for the sample of Haynes 282 in the initial state.

Table 1. Volume fraction of the main texture components in the sample of Haynes 282 in the initial state.

Texture Component {h k l} <u v w>	Volume Fraction [%]
(4 1 1) <2 5 $\bar{3}$>	5.0
Background	95.0

3.2.2. Texture of Samples of Haynes 282 Nickel Alloy Produced Using the DMLS Technique (in the Y-Z Sample Building Plane)

The measurements of texture after the printing process were carried out in the Y-Z sample building plane. The incremental printing of the alloy under study resulted in a sharp (clear) texture in 2/3 of the volume, with two components (Figures 6 and 7, Table 2). In more than half of the volume of the samples investigated, a cubic texture formed (0 0 1) <1 0 0>, as is characteristic of printed materials. In turn, in about 10% of the volume of the material a texture component was identified that is also present in the non-printed material (Figure 8)—in the as-delivered (reference) state, that is (4 1 1) <2 5 $\bar{3}$>.

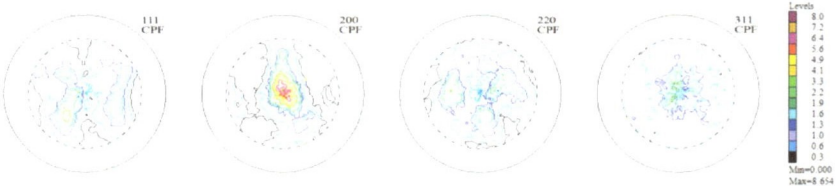

Figure 6. Experimental, incomplete pole figures (1 1 1), (2 0 0), (2 2 0), and (3 1 1) for the sample of Haynes 282 produced using DMLS.

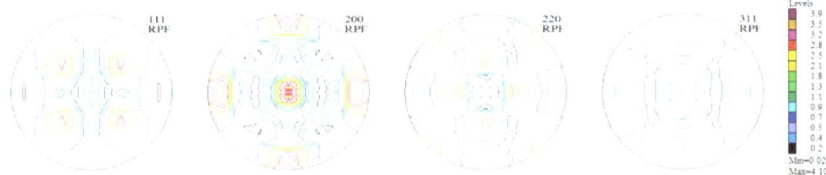

Figure 7. Complete pole figures (1 1 1), (2 0 0), (2 2 0), and (3 1 1) for the sample of Haynes 282 produced using DMLS.

Table 2. Volume fraction of the main texture components in the sample of Haynes 282 produced using DMLS.

Texture Component {h k l} <u v w>	Volume Fraction [%]
(0 0 1) <1 0 0>	54.0
(4 1 1) <2 5 $\bar{3}$>	10.0
Background	36.0

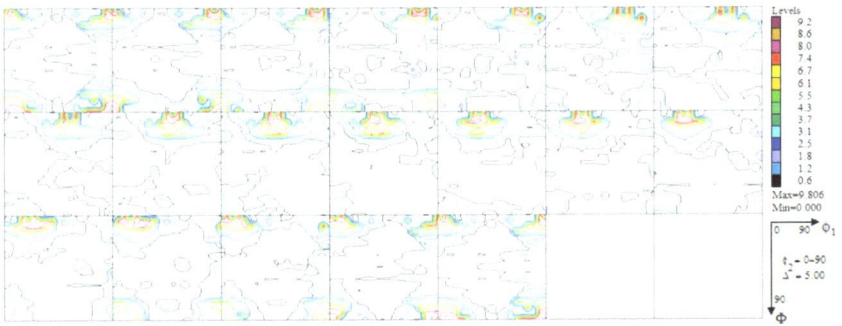

Figure 8. Orientation distribution function maps for the sample of Haynes 282 produced using DMLS.

3.2.3. Texture of Samples of Haynes 282 Nickel Alloy Produced Using the DMLS Technique (in the Y-Z Sample Building Plane) and after the Process of Aluminization

A significant reconstruction of the texture took place as a result of the thermo-chemical treatment applied to the printed material (the tests were conducted on the substrate without the layer). The texturing extended throughout almost the whole volume of the material (only about 7% of the volume was occupied by grains having a random orientation—Figure 9). Among the six texture components identified, two took up half the volume: ~(4 3 1) <3 $\bar{4}$ 0> and ~(1 5 3) <0 $\bar{3}$ 5>, around (0 1 1) <0 1 1> (~12% vol.). In short, one may acknowledge that almost 2/3 of the volume of the material accepted an orientation focused around (h k l) <1 1 0>. An exact analysis of the texture components as they emerged after the aluminization process (Figure 10, Table 3) showed that a large fraction of the volume of the material (25%) was occupied by grains dispersed around the orientation (h k l) <2 1 0>—the orientations (0 0 1) <2 $\bar{1}$ 0> and (1 0 2) <$\bar{2}$ 0 1>. This means that in almost all of the volume of the material produced using DMLS and subjected to the process of aluminization, grains dominated, whose orientations were dispersed around the components (h k l) <1 1 0> and (h k l) <2 1 0>, which is characteristic of textures after recrystallization (see Figure 11).

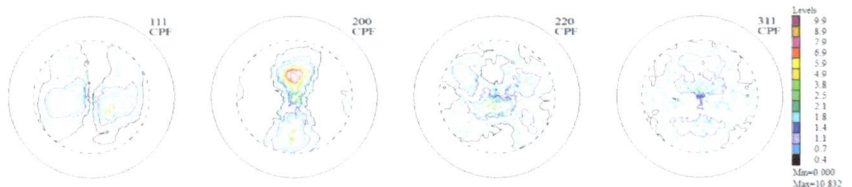

Figure 9. Experimental, incomplete pole figures (1 1 1), (2 0 0), (2 2 0), and (3 1 1) for the substrate of the sample of Haynes 282 produced using DMLS and after the process of aluminization (CVD).

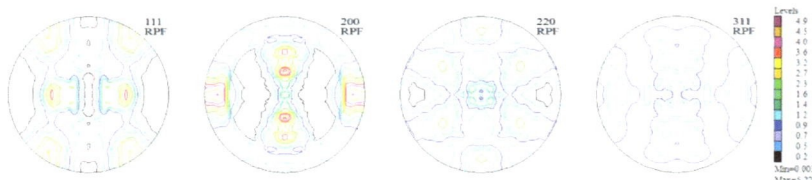

Figure 10. Complete pole figures (1 1 1), (2 0 0), (2 2 0), and (3 1 1) for the substrate of the sample of Haynes 282 produced using DMLS and after the process of aluminization (CVD).

Table 3. Volume fraction of the main texture components in the sample of Haynes 282 nickel alloy produced using DMLS and after the process of aluminization (CVD).

Texture Component {h k l} <u v w>	Volume Fraction [%]
(4 3 1) <3 $\bar{4}$ 0>	27.0
(1 5 3) <0 $\bar{3}$ 5>	23.0
(0 0 1) <2 $\bar{1}$ 0>	14.0
(0 1 1) <0 1 1>	12.0
(1 0 2) <$\bar{2}$ 0 1>	11.0
(4 $\bar{1}$ 1) <2 5 3>	6.0
Background	7.0

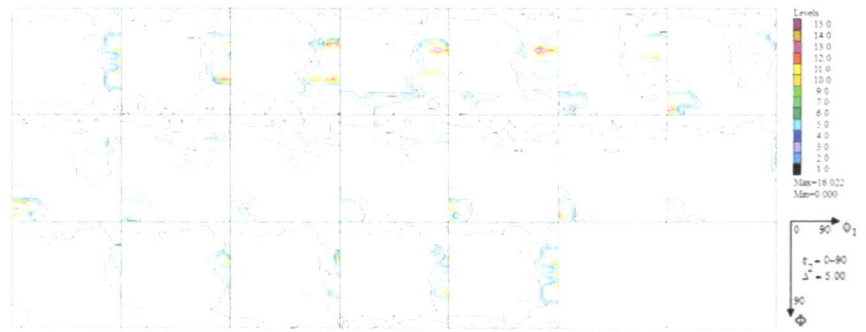

Figure 11. Orientation distribution functions for the substrate of the sample of Haynes 282 produced using DMLS and after the process of aluminization (CVD).

4. Conclusions

The utilization of DMLS (direct metal laser sintering) with Haynes 282 nickel alloy powder enables the production of samples with exceptional qualities, including a fine-crystalline structure and minimal porosity. This demonstrates the viability of DMLS as a reliable technique for fabricating high-quality metal components.

After the additive manufacturing process of the Haynes 282 alloy, a texture similar to that of the cast materials developed. In the printed material, however, a fine-crystalline structure was obtained. This underlines the importance of 3D printing in terms of optimizing the microstructure of materials, which leads to an improvement in their performance properties.

The process of aluminizing the Haynes 282 alloy produced by the DMLS technique leads to the transformation of the casting-type texture (initial state before the CVD process) to the recrystallization characteristic after the layer deposition process. This transformation is primarily attributed to the elevated temperatures associated with the CVD process, which leads to recrystallization in the substrate.

Author Contributions: Conceptualization, R.S. and J.M.; methodology, R.S. and J.M.; formal analysis, P.W., J.M., B.A.-C. and M.D.; investigation, R.S., B.A.-C. and J.M.; resources, J.M. and R.S.; data curation, J.M., P.M. and R.S.; writing—original draft preparation, R.S., J.M. and P.M.; writing—review and editing, R.S., P.W. and J.M.; visualization, J.M. and M.D.; supervision, R.S. and J.M.; project administration, R.S.; funding acquisition, R.S. All authors have read and agreed to the published version of the manuscript.

Funding: The research was funded by POB Technologie Materiałowe of Warsaw University of Technology in the framework of the Excellence Initiative: Research University (IDUB) programme.

Institutional Review Board Statement: Not applicable.

Informed Consent Statement: Not applicable.

Data Availability Statement: Data are available from the first author and can be shared with anyone upon reasonable request.

Conflicts of Interest: The authors declare no conflict of interest.

References

1. Kirka, M.M.; Unocic, K.A.; Kruger, K.; Forsythe, A. *Process Development for Haynes® 282® Using Additive Manufacturing*; Oak Ridge National Laboratory (ORNL): Oak Ridge, TN, USA, 2018. [CrossRef]
2. Vattappara, K.; Hosseini, V.A.; Joseph, C.; Hanning, F.; Andersson, J. Physical and thermodynamic simulations of gamma-prime precipitation in Haynes® 282® using arc heat treatment. *J. Alloys Compd.* **2021**, *870*, 159484. [CrossRef]
3. Ko, Y.S.; Kim, B.K.; Jung, W.-S.; Han, H.N.; Kim, D.-I. Effect of the microstructure of Haynes 282 nickel-based superalloys on oxidation behavior under oxy-fuel combustion conditions. *Corros. Sci.* **2022**, *198*, 110110. [CrossRef]
4. Alloy, N.; Tian, Z.; Zhang, C.; Wang, D.; Liu, W.; Fang, X. A Review on Laser Powder Bed Fusion of Inconel 625 Nickel-Based Alloy. *Appl. Sci.* **2020**, *10*, 81.
5. Akhtar, S.; Saad, M.; Misbah, M.R.; Sati, M.C. Recent advancements in powder metallurgy: A review. *Mater. Today Proc.* **2018**, *5*, 18649–18655. [CrossRef]
6. Druzgalski, C.L.; Ashby, A.; Guss, G.; King, W.E.; Roehling, T.T.; Matthews, M.J. Process optimization of complex geometries using feed forward control for laser powder bed fusion additive manufacturing. *Addit. Manuf.* **2020**, *34*, 101169. [CrossRef]
7. Kumar, S. Selective Laser Sintering/Melting. In *Comprehensive Materials Processing*; Elsevier: Amsterdam, The Netherlands, 2014; pp. 93–134. [CrossRef]
8. Cabrini, M.; Lorenzi, S.; Testa, C.; Pastore, T.; Brevi, F.; Biamino, S.; Fino, P.; Manfredi, D.; Marchese, G.; Calignano, F.; et al. Evaluation of Corrosion Resistance of Alloy 625 Obtained by Laser Powder Bed Fusion. *J. Electrochem. Soc.* **2019**, *166*, C3399–C3408. [CrossRef]
9. Singh, S.; Andersson, J. Heat-affected-zone liquation cracking in welded cast haynes® 282®. *Metals* **2020**, *10*, 29. [CrossRef]
10. Marchese, G.; Lorusso, M.; Parizia, S.; Bassini, E.; Lee, J.-W.; Calignano, F.; Manfredi, D.; Terner, M.; Hong, H.-U.; Ugues, D.; et al. Influence of heat treatments on microstructure evolution and mechanical properties of Inconel 625 processed by laser powder bed fusion. *Mater. Sci. Eng. A* **2018**, *729*, 64–75. [CrossRef]
11. Jackson, M.P.; Reed, R.C. Heat treatment of UDIMET 720Li: The effect of microstructure on properties. *Mater. Sci. Eng. A* **1999**, *259*, 85–97. [CrossRef]
12. Osoba, L.O.; Khan, A.K.; Adeosun, S.O. Cracking susceptibility after post-weld heat treatment in haynes 282 nickel based superalloy. *Acta Metall. Sin.* **2013**, *26*, 747–753. [CrossRef]
13. Li, C.; White, R.; Fang, X.Y.; Weaver, M.; Guo, Y.B. Microstructure evolution characteristics of Inconel 625 alloy from selective laser melting to heat treatment. *Mater. Sci. Eng. A* **2017**, *705*, 20–31. [CrossRef]
14. Deshpande, A.; Nath, S.D.; Atre, S.; Hsu, K. Effect of post processing heat treatment routes on microstructure and mechanical property evolution of haynes 282 Ni-based superalloy fabricated with selective laser melting (SLM). *Metals* **2020**, *10*, 629. [CrossRef]
15. Song, K.H.; Nakata, K. Effect of precipitation on post-heat-treated Inconel 625 alloy after friction stir welding. *Mater. Des.* **2010**, *31*, 2942–2947. [CrossRef]

16. Sitek, R.; Kaminski, J.; Mizera, J. Corrosion Resistance of the Inconel 740H Nickel Alloy after Pulse Plasma Nitriding at a Frequency of 10 kHz. *Acta Phys. Pol. A* **2016**, *129*, 584–587. [CrossRef]
17. Sitek, R. Influence of the high-temperature aluminizing process on the microstructure and corrosion resistance of the IN 740H nickel superalloy. *Vacuum* **2019**, *167*, 554–563. [CrossRef]
18. Sitek, R.; Molak, R.; Zdunek, J.; Bazarnik, P.; Wiśniewski, P.; Kubiak, K.; Mizera, J. Influence of an aluminizing process on the microstructure and tensile strength of the nickel superalloy IN 718 produced by the Selective Laser Melting. *Vacuum* **2021**, *186*, 110041. [CrossRef]
19. Romanowska, J. Aluminum diffusion in aluminide coatings deposited by the CVD method on pure nickel. *CALPHAD Comput. Coupling Phase Diagr. Thermochem.* **2014**, *44*, 114–118. [CrossRef]
20. Yavorska, M.; Sieniawski, J.; Zielińska, M. Functional properties of aluminide layer deposited on INCONEL 713 LC Ni-based super alloy in the CVD process. *Arch. Metall. Mater.* **2011**, *56*, 187–192. [CrossRef]

Disclaimer/Publisher's Note: The statements, opinions and data contained in all publications are solely those of the individual author(s) and contributor(s) and not of MDPI and/or the editor(s). MDPI and/or the editor(s) disclaim responsibility for any injury to people or property resulting from any ideas, methods, instructions or products referred to in the content.

Article

Thermal Barrier Stability and Wear Behavior of CVD Deposited Aluminide Coatings for MAR 247 Nickel Superalloy

Dominik Kukla [1],*, Mateusz Kopec [1], Zbigniew L. Kowalewski [1], Denis J. Politis [2], Stanisław Jóźwiak [3] and Cezary Senderowski [4]

[1] Institute of Fundamental Technological Research, Polish Academy of Sciences, Pawińskiego 5B, 02-106 Warsaw, Poland; mkopec@ippt.pan.pl (M.K.); zkowalew@ippt.pan.pl (Z.L.K.)
[2] Department of Mechanical and Manufacturing Engineering, University of Cyprus, 20537 Nicosia, Cyprus; politis.denis@ucy.ac.cy
[3] Faculty of Advanced Technologies and Chemistry, Military University of Technology, 00-908 Warsaw, Poland; stanislaw.jozwiak@wat.edu.pl
[4] Department of Materials and Machinery Technology, University of Warmia and Mazury, Oczapowskiego 11 St., 10-719 Olsztyn, Poland; cezary.senderowski@uwm.edu.pl
* Correspondence: dkukla@ippt.pan.pl

Received: 20 July 2020; Accepted: 31 August 2020; Published: 1 September 2020

Abstract: In this paper, aluminide coatings of various thicknesses and microstructural uniformity obtained using chemical vapor deposition (CVD) were studied in detail. The optimized CVD process parameters of 1040 °C for 12 h in a protective hydrogen atmosphere enabled the production of high density and porosity-free aluminide coatings. These coatings were characterized by beneficial mechanical features including thermal stability, wear resistance and good adhesion strength to MAR 247 nickel superalloy substrate. The microstructure of the coating was characterized through scanning electron microscopy (SEM), X-ray energy dispersive spectroscopy (EDS) and X-ray diffraction (XRD) analysis. Mechanical properties and wear resistance of aluminide coatings were examined using microhardness, scratch test and standardized wear tests, respectively. Intermetallic phases from the Ni-Al system at specific thicknesses (20–30 μm), and the chemical and phase composition were successfully evaluated at optimized CVD process parameters. The optimization of the CVD process was verified to offer high performance coating properties including improved heat, adhesion and abrasion resistance.

Keywords: chemical vapor deposition; nickel alloys; coatings; X-ray analysis

1. Introduction

Nickel alloys used for the construction of aircraft engines are characterized by high performance properties including corrosion, heat and creep resistance [1,2]. Additional aluminide coatings deposited on these alloys improve their thermal and chemical resistance during operation in high temperature and aggressive environments [3,4]. Specifically, aluminide thermal coatings have been found to effectively prevent oxidation and carbonization in high temperature conditions [5,6].

Increasing demands by the aircraft industry have resulted in improved engine efficiency and high operating temperatures [7,8], which has resulted in increased use of nickel alloys. Turbine components made from nickel alloys are exposed to extremely high temperatures, as the gas temperature upstream of the turbine may exceed 1650 °C [9]. In order to increase the operational life of turbine blades and other engine components, protective layers are commonly used. Thermal barrier coatings/bond coating (TBC/BC), with a complex chemical composition and structure, are currently used for turbine blades made of nickel alloys [10]. The outer ceramic layer is designed to reduce the thermal and erosive effects during take-offs and landings. Moreover, low thermal conductivity enables the elimination of rapid temperature changes and thermal expansion effects at different stages of engine operation. The bonding interlayer (bond coating, BC) is designed to protect engine components from oxidation and subsequently increases the adhesion of the ceramic layer to the substrate material. Such coatings are currently produced by plasma spraying (PS), electron-beam physical vapor deposition (EB PVD) and gas detonation spraying (GDS) techniques [11,12]. In plasma spraying, a powder of a carefully selected chemical composition is used to obtain the protective coating. In gas detonation spraying (GDS), FeAl intermetallic coatings obtained from self-decomposing powders are characterized by improved mechanical properties, such as hardness, thermal resistance and adhesive strength of the coating/substrate bond. The physical-chemical properties of the GDS protective interlayers with compact and lamellar structure could enhance the properties of the coating systems. It is possible to tailor the substrate coating by applying multiple structures and properties at individual areas of a workpiece, and thus control the influence of temperature and stress gradients. The application of an intermediate layer could improve the adhesion of GDS coatings through the enhancement of cohesive strength on grain boundaries. Moreover, the refinement and homogenization of the coating consisting of equiaxed, fine grains (<1 μm) can result in the formation of a multiphase, refractory intermetallic-based thermal barrier reinforced with Al_2O_3 oxides. Such intermediate layers could improve the adhesion between the interlayer and substrate by over 50 MPa. The complex microstructures of these layers also exhibit a uniform distribution of compressive stress throughout their volume, which has been confirmed by the Sachs–Davidenkov method [13]. Uniform stress distribution is typical for GDS processes, where there is limited thermal interaction between the detonation products and substrate. It may be concluded that the deposition of intermetallic coatings leads to thermal stress generation (due to differing linear thermal expansion coefficients), which can influence the integrity and durability of the coating/substrate bond system.

Therefore, the deposition of aluminum-based coatings must be performed with optimized process parameters to ensure the quality of the coating. Aluminide thermal coatings promise to reduce the sliding wear of turbine blades, and subsequently, increase the efficiency and lifetime of the engine. Taking into account the potential benefits for intermediate layer applications, the aim of this work was to determine the microstructural and mechanical properties of aluminide coatings deposited on a MAR 247 nickel superalloy using optimized chemical vapor deposition (CVD) process conditions in order to obtain a stable thermal barrier and wear resistant coating with satisfactory bond strength.

2. Materials and Methods

MAR 247 nickel superalloy was produced by using the casting process with uniform crystallization performed in ceramic molds. Sample casting was performed in an ALD vacuum furnace. Samples made of MAR 247 alloy with directional grain orientation (Figure 1a) were transferred out of the furnace at the controlled speed of 3 mm/min. Samples with equiaxed microstructure (EQ) were quenched in the furnace to achieve the required microstructure. The chemical composition of MAR 247 nickel superalloy is presented in Table 1.

Aluminide coatings were produced in the process of low activity aluminization by chemical vapor deposition (CVD) using an Ion-Bond setup (Figure 1b) (Ion Bond Bernex BPX Pro 325 S, IHI Ionbond AG, Olten, Switzerland) located in the Materials Testing Laboratory of the Rzeszów University of Technology. The CVD processes were performed on 50 samples of MAR 247 alloy in nitrogen/hydrogen protective atmospheres for 1–12 h at the temperature of 1040 °C and internal pressure of 150 mbar. The microstructural characterization and chemical composition analysis of the coatings were examined using a HITACHI SU70 scanning electron microscope (Hitachi, Tokyo, Japan) with energy dispersive spectroscopy attachment (EDS, by Oxford Instruments, Oxford, UK) and HITACHI 260 also with an EDS detector. Phase composition was evaluated using CuKα radiation with a Bruker D8 X-ray diffractometer (Bruker, Billerica, MA, USA) and Rigaku (Tokyo, Japan) Ultima IV diffractometer with Co-K radiation ($\lambda^1/41.78897$ Å) and operating parameters of 40 mA and 40 kV with a scanning speed of 1°/min and a scanning step of 0.02° in the range of 20°–150°. The microhardness of coatings was determined on a ZWICK hardness tester (Materialprüfung 3212002, Ulm, Germany) using the Vickers method. The sliding wear test was performed using the three rollers + cone method in accordance with the Polish Standard (PN-83/H04302). The tests were performed with the counter cone-sample made from SW7M high-speed steel under a constant pressure of 200 MPa and controlled time of 100 min. The adhesion of the coatings was assessed by applying a Micro-Combi-Tester (MCT[3], Anton Paar, Warsaw, Poland), with the force increasing from 0 to 100 N.

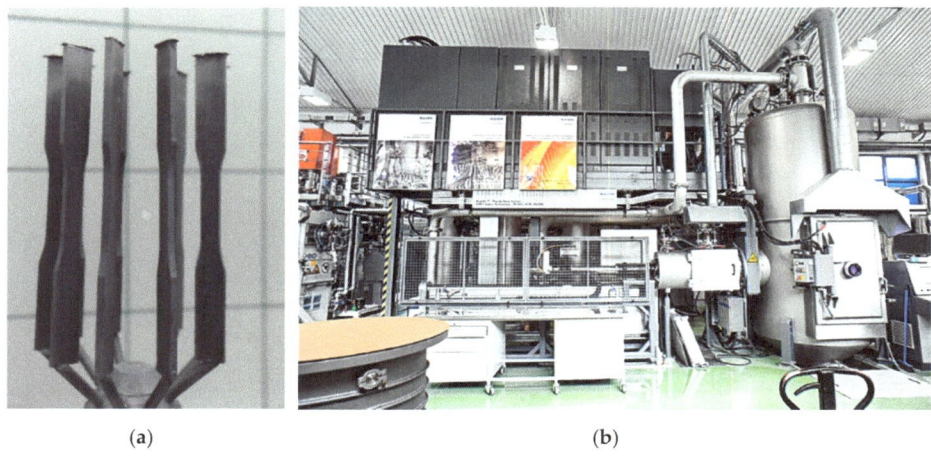

Figure 1. General view of MAR 247 cast samples (**a**), Ion Bond Bernex BPX Pro 325 S system for chemical vapor deposition (**b**).

Table 1. Chemical composition of MAR 247 nickel superalloy (wt.%).

C	Cr	Mn	Si	W	Co	Al	Ni
0.09	8.80	0.10	0.25	9.70	9.50	5.70	bal.

3. Results and Discussion

3.1. Microstructural Characterization of Coatings after CVD Process

The test matrix of the CVD parameters used in the study is presented in Table 2. The parameters include a temperature of 1040 °C with differing deposition times ranging from 1–12 h to assess the maximum effectiveness of the process. These parameters were selected on the basis of preliminary studies performed at different temperatures from the range of 880 °C to 1040 °C. In these tests, it was found that CVD performed at lower temperatures (880 °C, 950 °C) with a relatively long deposition time (up to 12 h) and with different protective atmospheres failed to obtain a defect free and cohesive coating, as presented in the sample cross-sections of Figure 2.

Table 2. The test matrix of CVD parameters.

Temperature [°C]	Deposition Time [h]	Protective Gas
1040	1	hydrogen
1040	2	hydrogen
1040	8	hydrogen
1040	12	hydrogen

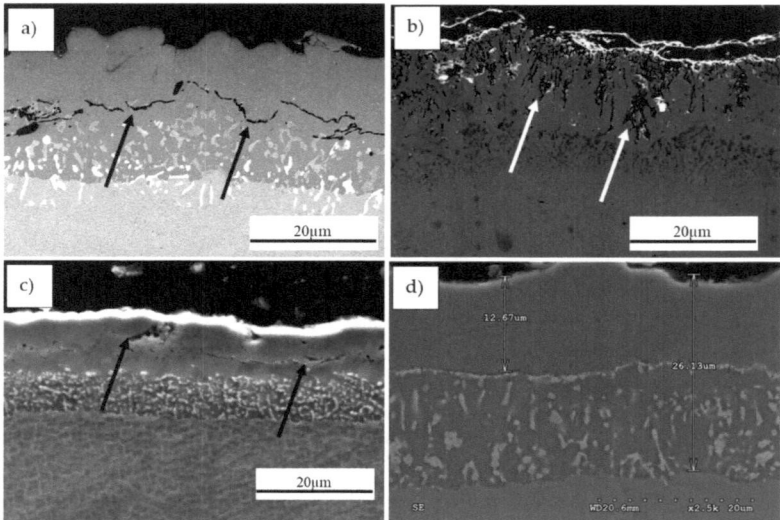

Figure 2. Cross-sections of intermetallic coatings produced by chemical vapor deposition (CVD) method on MAR 247 nickel superalloy at: (**a**) 880 °C for 12 h under nitrogen protective atmosphere; (**b**) 1040 °C for 8 h under nitrogen protective atmosphere; (**c**) 950 °C for 12 h under hydrogen protective atmosphere; (**d**) 1040 °C for 12 h under hydrogen protective atmosphere. Failure regions are indicated by arrows.

The deposition process performed at the initial temperature of 880 °C for 12 h under nitrogen protective atmosphere resulted in an incoherent coating with visible cracks along the surface (Figure 2a). Subsequent increase in the process temperature to 1040 °C led to excessive deterioration of the coating as shown in Figure 2b. Based on such observations, it was concluded that extension of the deposition time for a nitrogen protective atmosphere would have yielded unsatisfactory results. Therefore, similar process conditions were evaluated for the hydrogen atmosphere. Experimentation showed that at temperatures lower than 950 °C, defects such as cracks occurred within the structure (Figure 2c). Based on the microstructural observations, it was found that the optimal conditions include a temperature of 1040 °C with hydrogen as the protective gas atmosphere to perform the CVD process. These variables demonstrated the formation of a successful aluminide coating (Figure 2d), and thus, the process variable that was further studied to improve coating quality was deposition time.

The coating quality was evaluated by evaluating the homogenous morphology as well as the intermetallic behaviour. Figure 3 shows the surface morphology evolution of the deposited coatings depending on CVD deposition time with temperature and protective atmosphere maintained as constant. Aluminide coatings obtained during 1 h deposition time (Figure 3a,b) were characterized by ~10 μm radius craters and microvoids observed over the entire surface. The occurrence of these voids, as well as sharp-like grains in the highly developed cellular coating structures (Figure 3b), may indicate that the deposition time was too short to attain a protective coating without defects. After the deposition time was increased to 2 h (Figure 3c,d), the size of the craters slightly increased and a smooth structure was achieved. Despite the longer time of deposition, a porous structure was still observed (Figure 3d). A considerable change of the morphology was obtained when the deposition time was extended up to 8 h (Figure 3e–h). The coating deposited on MAR 247 (Figure 4a) after 12 h was characterized by a homogenous, non-defect structure (Figure 3g,h). Large, NiAl intermetallic crystallites were observed on the surface of the layer and their composition was confirmed by X-ray phase analysis (Figure 4b). The first peak observed on X-ray diffraction patterns (NiAl (100)) may indicate that formation of NiAl intermetallic superstructure (secondary solid solution β with B2 ordered structure) occurred. Such structure is stable at temperatures up to approximately 1400 °C with a wide range of aluminum content (31 at.% to 58 at.%) [14]. The chemical composition of this specific coating contained approximately 42.55% aluminum and 51.51% nickel, which may indicate that the NiAl was the dominant phase (Table 3).

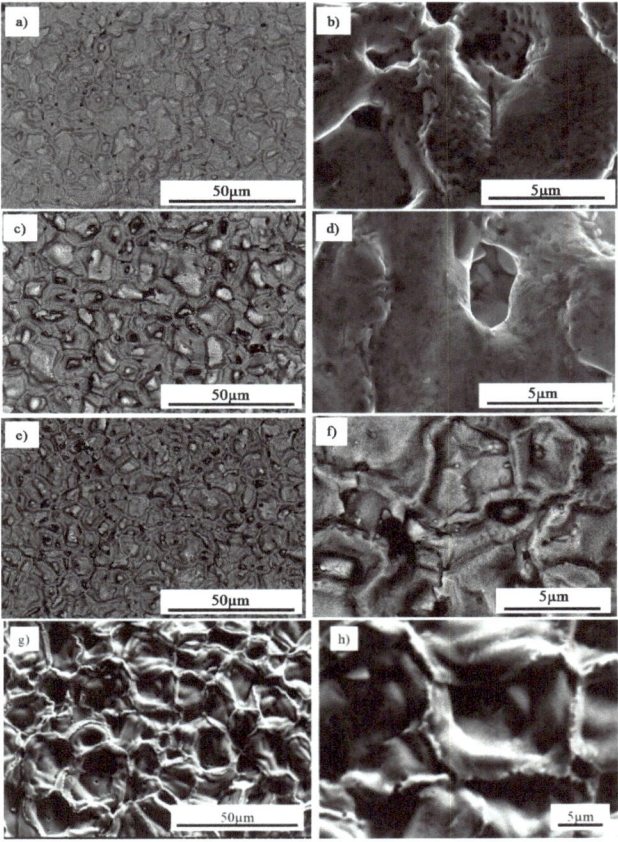

Figure 3. Microstructure of intermetallic coating produced by CVD method on MAR 247 nickel superalloy at 1040 °C for: (**a**,**b**) 1 h; (**c**,**d**) 2 h; (**e**,**f**) 8 h; (**g**,**h**) 12 h under hydrogen protective atmosphere.

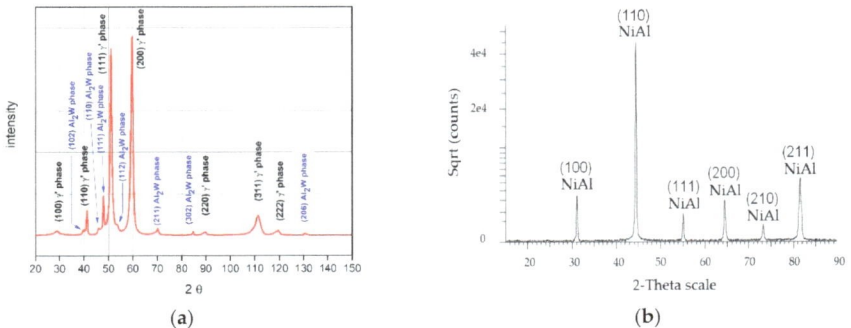

Figure 4. X-ray diffraction patterns from: (**a**) the as-received MAR 247 alloy without coating; (**b**) the coating produced by CVD process at the temperature of 1040 °C during 12 h of deposition under hydrogen protective atmosphere.

Table 3. Chemical EDS area composition of coating surface deposited at 1040 °C for 12 h under hydrogen protective atmosphere.

	Al	Cr	Fe	Co	Ni
at.%	42.55	0.66	0.43	4.85	51.51

Figure 5 presents an example of the intermetallic coating produced during the process of low-active high temperature aluminization (LAHT) at the temperature of 1040 °C for 12 h with additional distribution of main elements. This coating consists of two main sublayers. A chemical composition tested on the cross-section in micro-areas showed that the outer sublayer with a thickness of approximately 35 µm contains mainly aluminum (33.2 at.%) and nickel (55.4 at.%). In the inner sublayer, chromium (18.7 at.%) and molybdenum (9.96 at.%) were additionally observed. It was concluded that the NiAl layer limits chromium inter-diffusion, which may explain its low content in the outer sublayer and its increased concentration in the inner sublayer. The solubility of this element in the NiAl phase is limited to several percent. The EDS map analysis (Figure 5) also confirmed the assumptions of LAHT aluminization (900 °C–1150 °C) reported in literature [15]. In this process, the aluminum is deposited on the surface at a reduced rate. This allows nickel atoms to simultaneously diffuse outward to the surface and form a β-NiAl surface layer. The high content of nickel atoms observed in the coating area (Figure 5) was associated with the low aluminum activity at the surface, which effectively holds the surface aluminum content close to 50 at.%.

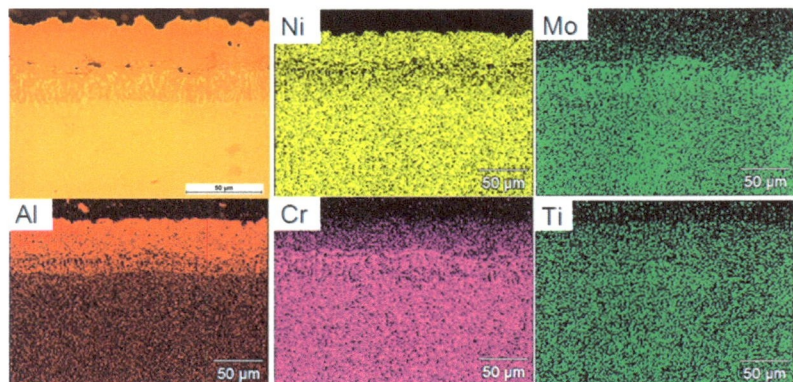

Figure 5. Analysis of the chemical composition from the coating surface after a low-active aluminization process at 1040 °C for 12 h.

Microstructural characterization and X-ray analysis allowed for the optimization of chemical vapor deposition (CVD) parameters of MAR 247 nickel superalloy. The CVD process performed at 1040 °C for 12 h in protective hydrogen atmosphere enabled a non-defect coating with homogenous structure and uniform thickness to be obtained (Figure 6).

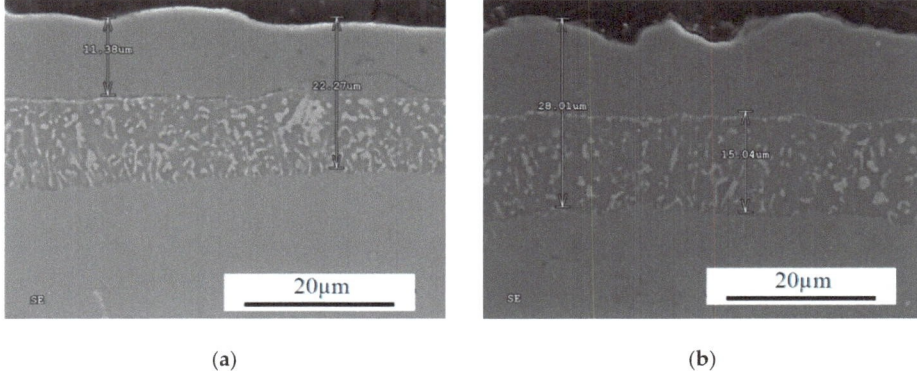

Figure 6. Cross-sections of aluminide coating with uniform thickness due to CVD process at the temperature of 1040 °C during 12 h of deposition. (**a**) the coating of 22.27 μm thickness; (**b**) the coating of 28.01 μm thickness.

3.2. Microhardness Profiles

Figure 7 shows the microhardness profile of aluminide coating. An evolution of the microhardness was captured from the edge of the aluminized sample to its core. It was clearly observed that the coating exhibits an improved microhardness in comparison to the substrate. The coating was characterized by hardness of 664 HV0.05, which was associated with the distribution of the aluminum atoms in the lattice of NiAl intermetallic coating. The enhancement in hardness could be also ascribed to the formation of NiAl phases during high temperature deposition. The hardness of the material gradually decreased from the edge to core of the material and maintained at the stable value approximately equal to 460 HV0.05. The maximum hardness, greater than 650 HV0.05, occurred in the coating area where Cr, Mo, Co carbides were dominant. The occurrence of such carbides reinforces the solution strengthening, and thus, improves the hardness. Furthermore, a high temperature promotes diffusion processes and reaction kinetics in the substrate/coating region, leading to the smooth hardness transition between coating areas, as shown in Figure 7. The microhardness of the transition zone was less than that of the aluminized coating surface because elements of the substrate (such as chromium and molybdenum) were excessively diffused during the process of solidification. Hardness in the transition area was approximately equal to 550 HV0.05, while that of the core of MAR 247 slightly increased to approximately 440 HV0.05. This is presumably caused by the relatively high temperature that the CVD process was executed in.

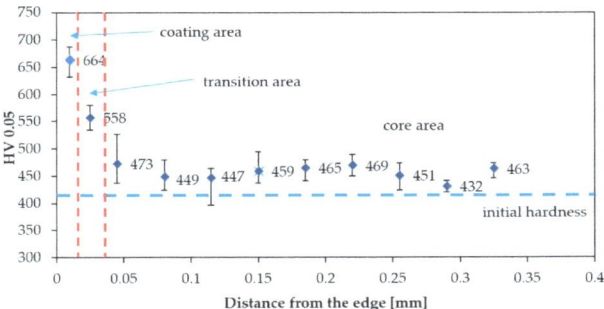

Figure 7. Distribution of microhardness in the cross-section of aluminized alloy.

3.3. Heat Resistance Properties of MAR 247 Alloy with Aluminide Protective Coating

As discussed in the introduction, modern aircraft engine turbines operate at high temperatures, which limits the materials that can be used. Contemporary nickel superalloys are limited to maximum operating temperatures of 1100 °C, which necessitates the use of protective coatings. Therefore, to replicate these operating conditions, heat resistance tests were performed on MAR 247 alloy with an aluminide protective coating in an air atmosphere at 1100 °C for 24 h.

The morphology of MAR 247 nickel superalloy in the initial state (without coating) (a), and with the NiAl diffusive intermetallic coating (b), are illustrated in Figure 8. X-ray phase analysis of the oxidized MAR 247 alloy without coating revealed that its phase structure contains mostly nickel oxide (NiO) (Figures 8a and 9a). On the contrary, the NiAl coated MAR 247 nickel superalloy exhibited no changes in the layer morphology. Subsequent SEM analysis revealed the fine-crystalline particles formed on the surface (Figure 8b). The phase structure of scale consisted mainly of NiAl and $NiAl_2O_4$ intermetallic phases as well as stable, alumina oxide (α-Al_2O_3) (Figure 9b). It was found that a high concentration of aluminum atoms near the coating surface area (Figure 5) allows for the formation of a thick, protective alumina oxide (α-Al_2O_3)-based scale and led to the improvement in hot corrosion resistance of the MAR 247 nickel superalloy. The aluminide coating subjected to oxidation was characterized by the excellent durability and tightness of the protective scale as no scale spallation was observed (Figure 8b).

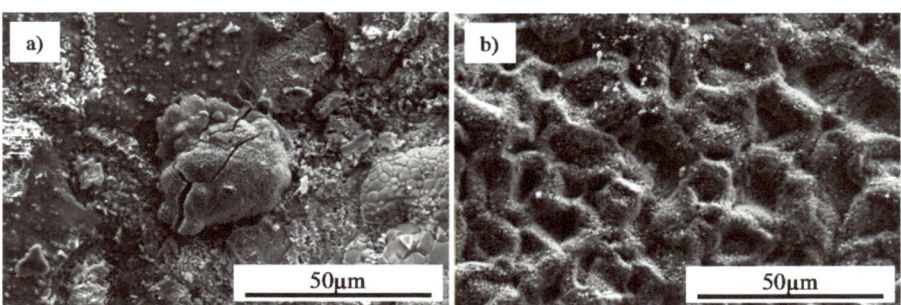

Figure 8. Morphology of the oxidized MAR 247 nickel superalloy: (**a**) in the initial state; (**b**) after the aluminization process.

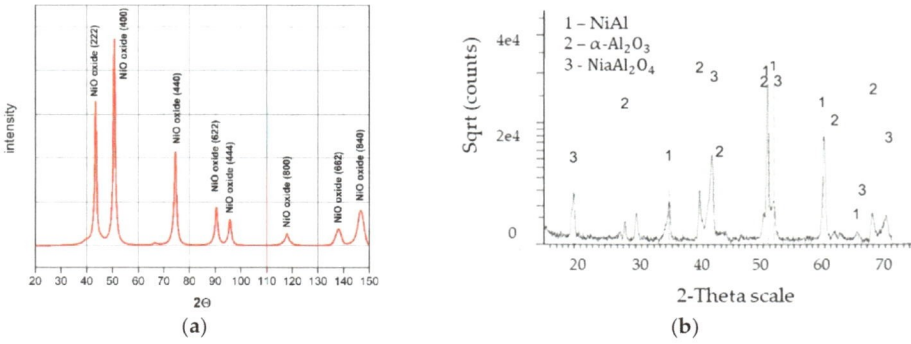

Figure 9. X-ray diffraction patterns determined for: (**a**) the raw MAR 247 alloy; (**b**) coated MAR 247 alloy after 24 h annealing in air conditions.

3.4. Characterization of Adhesion and Wear Resistance of MAR 247 Alloy with Aluminide Protective Coating

In order to assess the adhesion of the coatings, scratch tests were performed under a critical force of 100 N and 10 mm displacement. Despite the number of conformal cracks in the perpendicular to scratch direction, neither spallation nor a breakdown of the layer was observed on the coating surface, as shown in Figure 10. It should be noted that the aluminide coatings exhibited a very good adherence.

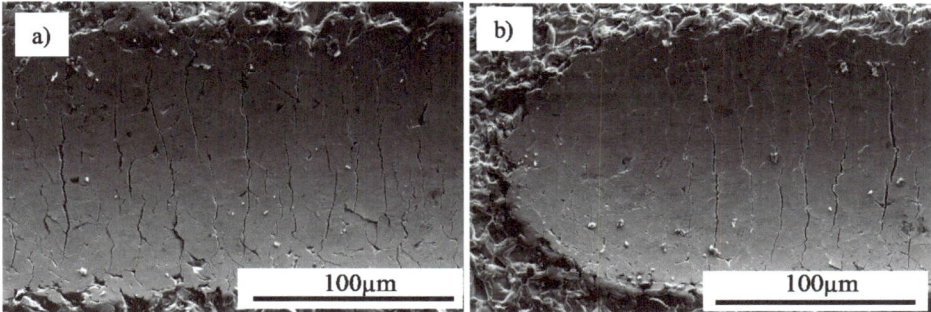

Figure 10. Morphology of the aluminide coating after scratch test: (**a**) central part and (**b**) scratch tip.

The results of wear resistance tests on the uncoated and coated MAR 247 alloy were presented in Figure 11. The uncoated MAR 247 alloy was characterized by the linear wear of 22 µm after 100 min of friction, which is twice as large as that for the coated alloy. A significant wear increase in the uncoated MAR 247 alloy can be easily observed after approximately 30 min. It was noted that after a relatively long time of friction (100 min), the NiAl coating remains firmly adhered to the substrate. The microstructural assessment of worn tracks (Figure 12) suggests the formation of third-body particles, associated with the observed scoring and grooving shallow marks. The presence of third-body particles in sliding systems is widely reported in the literature [16,17], and is related to the detachment of material from the surfaces being in the contact during sliding.

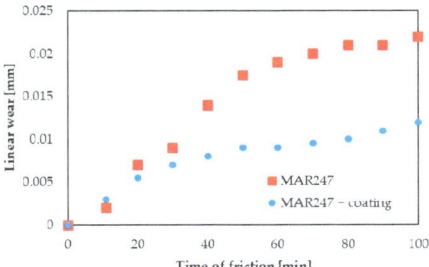

Figure 11. Linear wear of the as-received MAR 247 alloy without and with protective coating.

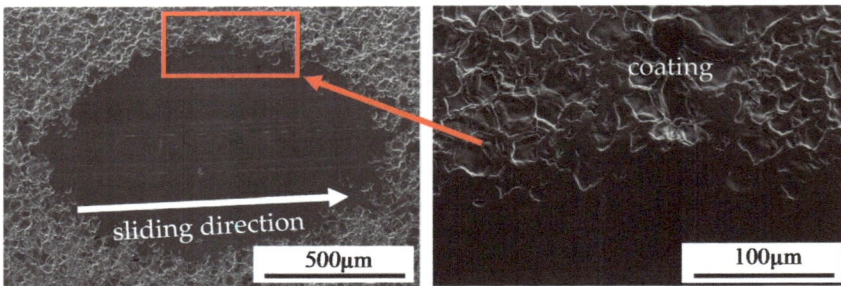

Figure 12. Marks of wear for MAR 247 nickel superalloy with NiAl coating.

4. Conclusions

Chemical vapor deposition for MAR 247 nickel superalloy was performed at a temperature of 1040 °C for 12 h in a protective hydrogen atmosphere. This process was demonstrated to produce a non-defect substrate material with thermal barrier and wear resistant uniform thickness NiAl coating from 20 µm to 30 µm. The coating was characterized by the very good adherence, wear and thermal resistance confirmed by experimental studies. Its application improved the mechanical properties, such as hardness and wear, by almost a factor of two compared to the as-received MAR 247 alloy. The coating hardness of 664 HV0.05 was associated with the distribution of aluminum atoms in the lattice of the NiAl intermetallic coating formed during high temperature deposition. The aluminide coatings exhibited a very good adherence during scratch tests as the breakdown of the layer was not observed on the coating surface after testing. The oxidized aluminide coating was characterized by the excellent durability and tightness of the protective scale, as no scale spallation was observed. The phase structure of the scale consisted mainly of NiAl and $NiAl_2O_4$ intermetallic phases as well as α-Al_2O_3 stable oxide that improved the hot corrosion resistance of nickel superalloys. It was thus demonstrated that the CVD technology with optimized parameters could be successfully applied to enhance the performance of nickel superalloys.

Author Contributions: Conceptualization, M.K. and D.K.; methodology, D.K., M.K. and S.J.; formal analysis, D.K., M.K., C.S., S.J. and Z.L.K.; writing—original draft preparation, M.K.; writing—review and editing, C.S., D.J.P. and Z.L.K. All authors have read and agreed to the published version of the manuscript.

Funding: The authors gratefully acknowledge the funding by The National Centre for Research and Development, Poland, under Program for Applied Research, grant no. 178781.

Acknowledgments: The authors are grateful to Mirosław Wyszkowski and Andrzej Chojnacki from the Institute of Fundamental Technological Research of the Polish Academy of Sciences for their great support during the experimental work.

Conflicts of Interest: The authors declare no conflict of interest.

References

1. Agarwal, D.C.; Brill, U. High-Temperature-Strength Nickel Alloy. *Adv. Mater. Process.* **2000**, *158*, 31–34.
2. Boutarek, N.; Saïdi, D.; Acheheb, M.A.; Iggui, M.; Bouterfaïa, S. Competition between three damaging mechanisms in the fractured surface of an Inconel 713 superalloy. *Mater. Charact.* **2008**, *59*, 951–956. [CrossRef]
3. Kalivodova, J.; Baxter, D.; Schutze, M.; Rohr, V. Corrosion behaviour of boiler steels, coatings and welds in flue gas environments. *Mater. Corros.* **2008**, *59*, 367–373. [CrossRef]
4. Kochmańska, A.; Garbiak, M. High-temperature diffusion barrier for Ni–Cr Cast Steel. *Defect Diffus. Forum* **2011**, *312*, 595–600. [CrossRef]
5. Zhan, Z.; He, Y.; Li, L.; Liu, H.; Dai, Y. Low-temperature formation and oxidation resistance of ultrafine aluminide coatings on Ni-base superalloy. *Surf. Coat. Technol.* **2009**, *203*, 2337–2342. [CrossRef]

6. Xu, Z.H.; Dai, J.W.; Niu, J.; He, L.M.; Mu, R.D.; Wang, Z.K. Isothermal oxidation and hot corrosion behaviors of diffusion aluminide coatings deposited by chemical vapor deposition. *J. Alloys Compd.* **2015**, *637*, 343–349. [CrossRef]
7. Goward, G.W. Progress in Coatings for Gas Turbine Airfoils. *Surf. Coat. Technol.* **1998**, *1*, 73–79. [CrossRef]
8. Barbosa, C.; Nascimento, J.L.; Caminha, I.M.V.; Abud, I.C. Microstructural aspects of the failure analysis of nickel base superalloys components. *Eng. Fail. Anal.* **2005**, *12*, 348–361. [CrossRef]
9. Moustapha, H.; Zeleski, M.F.; Baines, N.C.; Japikse, D. *Axial and Radial Turbines*; Concepts NREC: White River Junction, VT, USA, 2003.
10. Tamarin, Y. *Protective Coatings for Turbine Blades*; ASM International: Materials Park, OH, USA, 2002.
11. Senderowski, C.; Bojar, Z. Gas detonation spray forming of Fe–Al coatings in the presence of interlayer. *Surf. Coat. Technol.* **2008**, *202*, 3538–3548. [CrossRef]
12. Senderowski, C.; Bojar, Z. Influence of detonation gun spraying conditions on the quality of Fe-Al intermetallic protective coatings in the presence of NiAl and NiCr interlayers. *J. Spray Technol.* **2009**, *18*, 435–447. [CrossRef]
13. Senderowski, C.; Bojar, Z.; Roy, G.; Czujko, T.; Wołczyński, W. Residual stresses determined by the modified Sachs method within a gas detonation sprayed coatings of the Fe-Al intermetallic. *Arch. Met. Mater.* **2007**, *52*, 569–578.
14. Chaplygina, A.A.; Chaplygin, P.A.; Starostenkov, M.D. Structural transformations in the NiAl alloys with deviations from the stoichiometric composition during stepwise cooling. *IOP Conf. Ser. Mater. Sci. Eng.* **2018**, *447*, 012054. [CrossRef]
15. Sequeira, C.A.C. *High Temperature Corrosion: Fundamentals and Engineering*; Willey: Hoboken, NJ, USA, 2019; ISBN 978-0-470-11988-4.
16. Inman, I.A.; Datta, P.K.; Du, H.L.; Burnell-Gray, J.S.; Pierzgalski, S.; Luo, Q. Studies of high temperature sliding wear of metallic dissimilar interfaces. *Tribol. Int.* **2005**, *38*, 812–823. [CrossRef]
17. Jiang, J.; Stott, F.H.; Stack, M.M. The role of triboparticulates in dry sliding wear. *Tribol. Int.* **1998**, *31*, 245–256. [CrossRef]

© 2020 by the authors. Licensee MDPI, Basel, Switzerland. This article is an open access article distributed under the terms and conditions of the Creative Commons Attribution (CC BY) license (http://creativecommons.org/licenses/by/4.0/).

Article

Nondestructive Methodology for Identification of Local Discontinuities in Aluminide Layer-Coated MAR 247 during Its Fatigue Performance

Dominik Kukla [1], Mateusz Kopec [1,2,*], Kehuan Wang [3], Cezary Senderowski [4,*] and Zbigniew L. Kowalewski [1]

[1] Institute of Fundamental Technological Research, Polish Academy of Sciences, Pawińskiego 5B, 02-106 Warsaw, Poland; dkukla@ippt.pan.pl (D.K.); zkowalew@ippt.pan.pl (Z.L.K.)
[2] Department of Mechanical Engineering, Imperial College London, London SW7 2AZ, UK
[3] State Key Laboratory of Advanced Welding and Joining, Harbin Institute of Technology, Harbin 150001, China; wangkehuan@hit.edu.cn
[4] Department of Materials Technology and Machinery, University of Warmia and Mazury, Oczapowskiego 11 St., 10-719 Olsztyn, Poland
* Correspondence: mkopec@ippt.pan.pl (M.K.); cezary.senderowski@uwm.edu.pl (C.S.)

Abstract: In this paper, the fatigue performance of the aluminide layer-coated and as-received MAR 247 nickel superalloy with three different initial microstructures (fine grain, coarse grain and column-structured grain) was monitored using nondestructive, eddy current methods. The aluminide layers of 20 and 40 µm were obtained through the chemical vapor deposition (CVD) process in the hydrogen protective atmosphere for 8 and 12 h at the temperature of 1040 °C and internal pressure of 150 mbar. A microstructure of MAR 247 nickel superalloy and the coating were characterized using light optical microscopy (LOM), scanning electron microscopy (SEM) and X-ray energy dispersive spectroscopy (EDS). It was found that fatigue performance was mainly driven by the initial microstructure of MAR 247 nickel superalloy and the thickness of the aluminide layer. Furthermore, the elaborated methodology allowed in situ eddy current measurements that enabled us to localize the area with potential crack initiation and its propagation during 60,000 loading cycles.

Keywords: chemical vapor deposition; nickel alloys; aluminide coatings; fatigue; eddy current

Citation: Kukla, D.; Kopec, M.; Wang, K.; Senderowski, C.; Kowalewski, Z.L. Nondestructive Methodology for Identification of Local Discontinuities in Aluminide Layer-Coated MAR 247 during Its Fatigue Performance. *Materials* **2021**, *14*, 3824. https://doi.org/10.3390/ma14143824

Academic Editor: Christian Motz

Received: 25 May 2021
Accepted: 5 July 2021
Published: 8 July 2021

Publisher's Note: MDPI stays neutral with regard to jurisdictional claims in published maps and institutional affiliations.

Copyright: © 2021 by the authors. Licensee MDPI, Basel, Switzerland. This article is an open access article distributed under the terms and conditions of the Creative Commons Attribution (CC BY) license (https://creativecommons.org/licenses/by/4.0/).

1. Introduction

Nickel superalloys are commonly used in aircraft engines due to their superior, high-temperature performance properties, including corrosion, heat and creep resistance [1,2]. The most conventional nickel superalloys for gas turbines are MAR 247 [3], Rene 80 [4] and IN738 [5]. It should be emphasized that MAR 247 exhibits higher strength properties and better creep response in comparison to conventional alloys for aircraft engine components [3–5]. Increased demand for new materials in the aircraft industry led to the fabrication of new generation, cost-effective alloys doped with Re and Ru [6]. However, high costs limit the application of new generation materials in mass production.

The fatigue life of nickel-based superalloys is mainly determined by their initial microstructure, morphology and volume fraction of γ' and γ'' precipitates [7]. In a comprehensive review presented by Garimella et al. [7], it was reported that ageing accompanied by the precipitation of γ' precipitates could enhance fatigue properties. It was shown that the crack-growth rates were lower for materials that had been aged to peak hardness compared to the underaged and as-received specimens. Moreover, it was further confirmed that the maximum stress increased with increasing volume of the γ' participation up to the peak-aged condition and decreased in the overaged condition. It should also be mentioned that materials with the coarse-grained structure exhibit a shorter low-cycle fatigue life than fine-grained ones [7]. Furthermore, finer precipitates resulted in a slower long-crack growth rate and longer low-cycle fatigue life (LCF) of the nickel-based superalloys, such as Inconel 718 [7].

In order to enhance the mechanical properties and service life of commonly used nickel superalloys, the Ni-Al-type intermetallics were used as coating materials. These materials are characterized by a crystalline structure with strong chemical bonds and tightly packed atoms in the lattice, which improve their thermal stability significantly. It was found that only NiAl and Ni_3Al coatings could form a beneficial coating structure that could transfer high mechanical loads in aggressive environments [8,9]. Ni_3Al-based intermetallics exhibited improved fatigue strength at high temperatures in comparison to the commonly used Ni-based superalloys [10]. The superior properties of thermally stable NiAl- and Ni_3Al-based materials at high temperatures could be potentially used as the protective coatings for aircraft engine turbines [11,12]. Generally, thermal barrier coatings are applied to the parts of a gas turbine to reduce their operational temperature by approximately 100–300 °C while simultaneously increasing their service life [13]. The content of aluminum and chromium within the coating allows it to form a stable oxide layer that protects the substrate material and significantly reduces its chemical reactivity during performance in aggressive environments [14–16]. Since a high temperature significantly reduces the fatigue life of nickel superalloys [16], the application of thermal barrier coatings could further extend their fatigue life. The high-temperature fatigue response of the coated MAR 247 nickel superalloy has been widely studied in the literature [17–20]. It should be mentioned, however, that before the engine attains the high working temperature, the turbine blades are subjected to loading at ambient temperature. On the other hand, the nickel-based superalloys used for aero application mainly work at elevated temperatures, and thus, most studies were devoted to testing in aggressive environments. However, before any part will start its high-temperature performance, it should withstand the high load at ambient temperatures up to 300 °C. Nickel-based superalloys exhibit microstructural and mechanical stability at such temperatures, and therefore, the fatigue testing performed at room temperature allows us to characterize their high cyclic load capability. Unfortunately, there are not many available publications devoted to the effect of coating on the fatigue performance of the nickel-based superalloys at ambient temperatures. Increasing demands within the aviation industry for a new generation of turbine blade materials led to the development of new diagnostic techniques. Advanced techniques, such as the blade-surface images analysis [21], optoelectronic and thermographic methods [22], eddy current and ultrasonic methods [23], vibrothermography [24] or even Digital Image Correlation [25], allow for the identification of cracks, intrinsic defects, subsurface defects, pores and potential areas of crack initiation. The variety of destructive and nondestructive methods for nickel-based superalloy inspection allows us to identify potential reasons for the material decohesion. However, a correlation between destructive and nondestructive methods has not been sufficiently reported as yet.

Therefore, the main aim of this work was to assess the effect of aluminide layer thickness and the initial microstructure of MAR 247 on its fatigue performance during in situ eddy current measurements. Such tests enabled the characterization of the mechanical properties of three different initial microstructures and two different thicknesses of NiAl coating obtained by using the CVD process with optimized parameters presented in the authors' previous paper [26] and assessed the effectiveness of the eddy current method for in situ measurements that allow the identification of areas of potential crack. The specific microstructures and coating thicknesses were used in this paper due to their superior high temperature performance reported by authors in a different paper [27].

2. Materials and Methods

Specimens made of MAR 247 nickel superalloy were manufactured using conventional casting process in an ALD vacuum furnace (ALD Vacuum Technologies GmbH, Hanau, Germany). MAR 247 nickel superalloy specimens with equiaxed microstructures (EQ) were quenched with the furnace to achieve the required fine (Figure 1a) and coarse (Figure 1b) microstructure. In order to classify the sizes of equiaxed grains, the cast specimens were thermally insulated using ceramic wool (3 layers for fine-grained and 6 layers for coarse-grained microstructures). Specimens with directional grain orientation (DS) (Figure 1c) were transferred outside the furnace under controlled speeds of 3 mm/min, which enabled the formation of the columnar structure. The chemical composition of MAR247 nickel superalloy was presented in Table 1.

(a)   (b)   (c)

Figure 1. Macroscopic structure of MAR 247 nickel superalloy: (**a**) fine grain, (**b**) coarse grain and (**c**) column grain observed by using light optical microscope.

Table 1. Chemical composition of MAR 247 superalloy (wt. %).

C	Cr	Mn	Si	W	Co	Al	Ni
0.09	8.80	0.10	0.25	9.70	9.50	5.70	bal.

Aluminide coatings were obtained using the CVD process and Ion-Bond setup (Ion Bond Bernex BPX Pro 325 S, IHI Ion bond AG, Olten, Switzerland) located in the Materials Testing Laboratory for the Aviation Industry of the Rzeszów University of Technology, Poland. The optimized CVD parameters were obtained for the hydrogen protective atmosphere, with a deposition time of 8 and 12 h at the temperature of 1040 °C and internal pressure of 150 mbar [27]. The microstructural observations and chemical composition analysis were performed using a Hitachi 2600N scanning electron microscope with an energy dispersive spectroscopy (EDS) attachment (Oxford Instruments, Oxford, UK). The microhardness of the as-received and coated material was determined on a ZWICK hardness tester (Materialprüfung 3212002, Ulm, Germany) using the Vickers method. Standard fatigue tests were performed using the MTS 810 testing machine (MTS System, MN, USA) and the conventional MTS extensometer. Uniaxial tensile tests were carried out at strain rate equal to 2×10^{-4} s^{-1} using five specimens. Fatigue tests at a temperature of 23 °C were force controlled under the zero mean level, constant stress amplitude and a frequency of 20 Hz. Every fatigue test was performed at least twice to guarantee the reliability of the results obtained. The range of stress amplitude from 350 to 650 MPa was established on the basis of the conventional yield point $R_{0.2}$ determined from the uniaxial tensile test. The geometry of the specimens is presented in Figure 2. The eddy current measurements were performed using an Olympus Nortec 600 D (Olympus, Tokio, Japan) flaw detector and pencil probes with operating frequencies of 100–500 kHz and 1–6 MHz. The signal was calibrated using a standardized pattern sample with reference electric discharge machined (EDM) notches of 0.1, 0.2, 0.5 and 1 mm depth. Such a sample enables the optimization of the measuring parameters for the best surface and subsurface defect detection. EC measurements were performed on three different areas located in the gauge length and on both sides of the specimen using frequency equal to 240 Hz, after each of 10,000 cycles. Selected values of frequency enabled a penetration for depth of approx. 1 mm for 240 Hz.

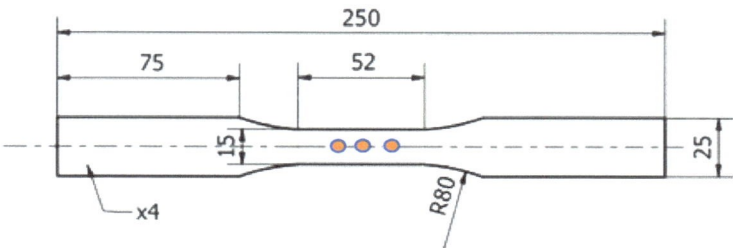

Figure 2. Geometry of the specimen used for uniaxial tensile and standard fatigue tests with marked areas of EC measurements (units in mm).

3. Results and Discussion

3.1. Microstructural Characterization of Coating during CVD Process

The microstructure of the MAR 247 nickel superalloy after the CVD process performed at 1040 °C revealed the presence of columnar and equiaxial dendrites, as shown in Figure 3. Based on the quenching conditions applied, different structures were observed. A fine-grained structure (Figure 3a) was achieved during slow cooling with the furnace. It should be mentioned that the static recrystallization temperature of the nickel-base alloys is within the range from 1000 to 1100 °C [28]. During the CVD process, the specimens were exposed to temperatures over 1000 °C, and therefore, a recrystallization could occur. During high-temperature exposure, the deformed and elongated grains would transform into finer equiaxial grains [29]. A coarse-grained structure (Figure 3b) was characterized by large grains, which may indicate that high-temperature exposure led to extensive grain growth. On the other hand, columnar grains (Figure 3c) grew parallel to the solidification direction. They were formed after the casting process, performed under carefully controlled speeds.

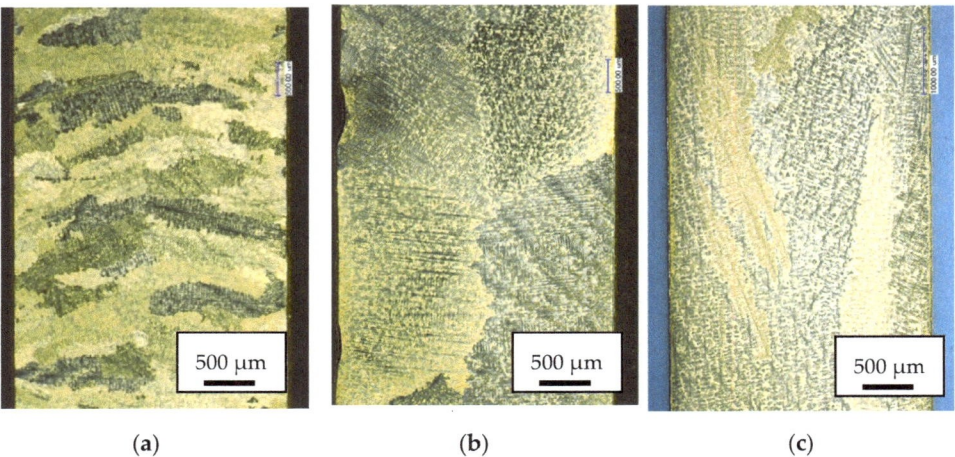

Figure 3. Microscopic structure of the MAR 247 nickel superalloy: (**a**) fine grain, (**b**) coarse grain and (**c**) column grain observed by using light optical microscope.

The specimen view of 40 µm-thick NiAl coating is presented in Figure 4. Its structure was mainly determined by the growth kinetics and conditioned by the temperature, pressure and different synthesis time in the CVD process. NiAl coating was uniformly distributed on the MAR 247 surface. The cross-sectional view allowed us to distinguish its two-layer structure: a homogeneous zone of secondary solid solution of the β (NiAl) phase and heterogeneous NiAl matrix (dark grey) with Ni_3Al phase dispersions (bright grey). The lower content of Al and the participation of Co, Cr and Ti alloying elements were found in the interlayer zone. This is due to the atom diffusion from the substrate. The chemical composition analysis was performed in the cross-section of specimens at seven points starting from the edge into the substrate (Figure 4, Table 2). The content of aluminum gradually decreased from 25% on the edge to ~6% within the substrate material. The coating was characterized by the typical intermetallic superstructure of the secondary solid solution β with B2 ordered structure [30]. One can conclude that the CVD process parameters were successfully selected since no defects were observed between the coated material and the coating itself. Minor cracks between sub-layers were caused by extensive grinding during the preparation of the metallographic specimens for the microstructural observations. The detailed microstructural characterization and mechanical properties of the NiAl coating obtained, including hot resistance, adhesion and wear resistance, were presented in the authors' previous work [26].

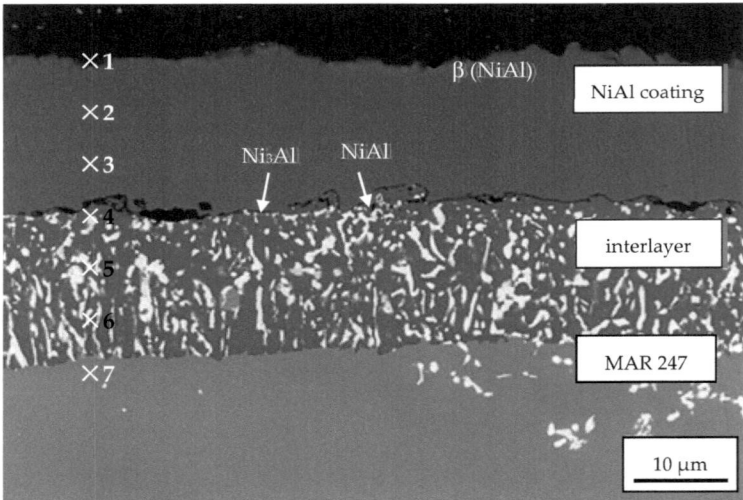

Figure 4. Microstructure of 40 µm NiAl coating produced by the CVD method on the MAR 247 nickel superalloy at 1040 °C with points of EDS analysis marked on crack propagation area.

Table 2. Chemical composition (wt. %) of coating surface obtained after deposition at 1040 °C.

Point	Al	Si	Ti	Cr	Co	Ni
1	25.01	x	x	2.66	7.78	64.55
2	23.68	x	x	3.23	7.95	65.15
3	21.79	x	0.49	4.63	8.69	64.41
4	19.31	0.34	0.89	6.52	9.57	63.00
5	18.73	0.51	1.07	5.95	9.84	63.64
6	17.41	0.53	1.20	7.07	9.44	63.98
7	6.47	0.23	0.65	7.21	9.71	59.85

3.2. Effect of Coating Thickness on Fatigue Behavior

The effect of coating thickness on the mechanical properties of the MAR247 nickel-based superalloy was also investigated in the standard fatigue tests. The results of experiments showed that the coating slightly decreased the stress response of the MAR 247 specimens at temperatures of 23 °C (Figure 5). Based on the analysis of S-N curves, it can be assumed that both fine- and coarse-grained microstructures do not affect the fatigue behavior of the MAR 247 nickel-based superalloy, since similar stress responses were achieved for both of these structures. It should be mentioned, however, that the column grain structure was characterized by the weakest properties in comparison to other structures. It was reported by Sulak et al. [31] that the cyclic deformation of nickel-based superalloys is mainly determined by the interaction between dislocations and γ strengthening phase. This is due to the fact that the precipitates played a dominant role as an effective barrier against the dislocation movement. Regardless of the initial microstructure, it was found that a difference in coating thickness affects the stress response of the MAR 247 nickel-based superalloy. A slight reduction in the stress amplitude could be observed when the thicker coating was applied to the substrate material. Such behavior might be caused by the high hardness and stiffness of the protective coating. Similar findings were presented in [32–34], where the fatigue life of steel decreased with increasing coating thickness. Such effect was explained using the linear elastic fracture mechanics under the assumption of small-scale yielding [33]. It was found that the strain energy stored in the coating layer contributed to the crack development. On the other hand, Akebono et al. [32,35] reported that two main factors could affect the fatigue performance of the Ni-based coated steel substrate. The first one was found to be the size and population of porosities in the coating. As the fatigue crack initiation was taking place at the porosities, specimens exhibited lower fatigue strength. The second factor was related to the lowered hardness of the Ni matrix accompanying chromium segregation. The same behavior can be observed in Figures 5–7, where the 40 μm-thick NiAl coating affected the fatigue performance more than that of the 20 μm one. It was found that fusion for a shorter time is more effective in producing sprayed materials of better fatigue properties [35].

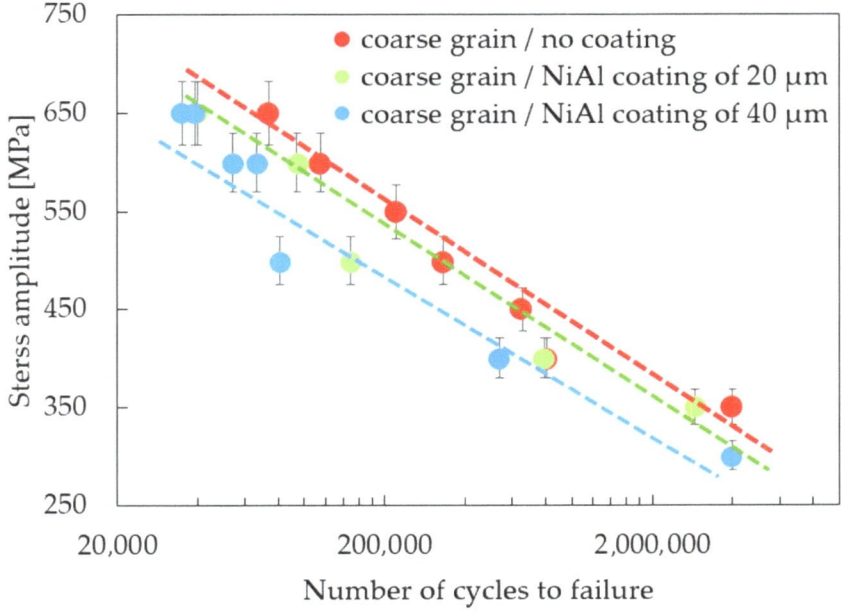

Figure 5. S-N curves for the coated, coarse-grained MAR 247 nickel-based superalloy.

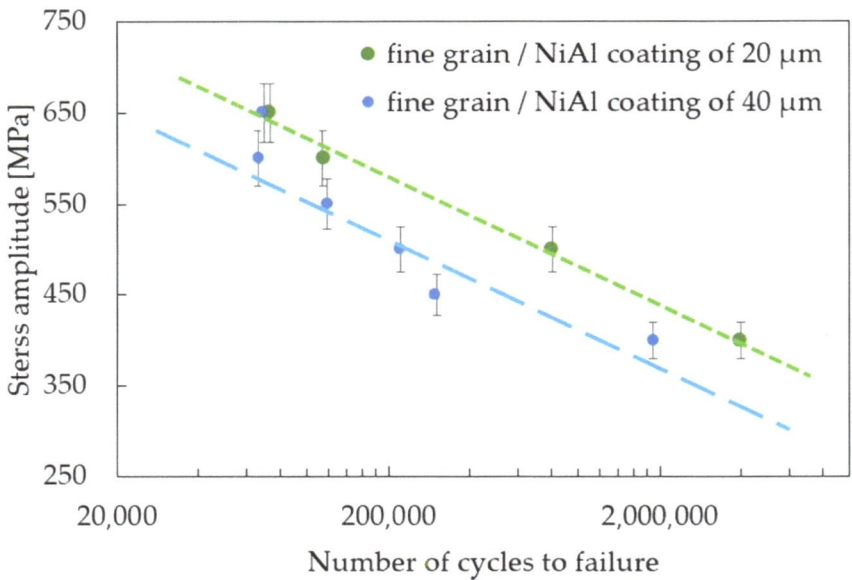

Figure 6. S-N curves for the coated, fine-grained MAR 247 nickel-based superalloy.

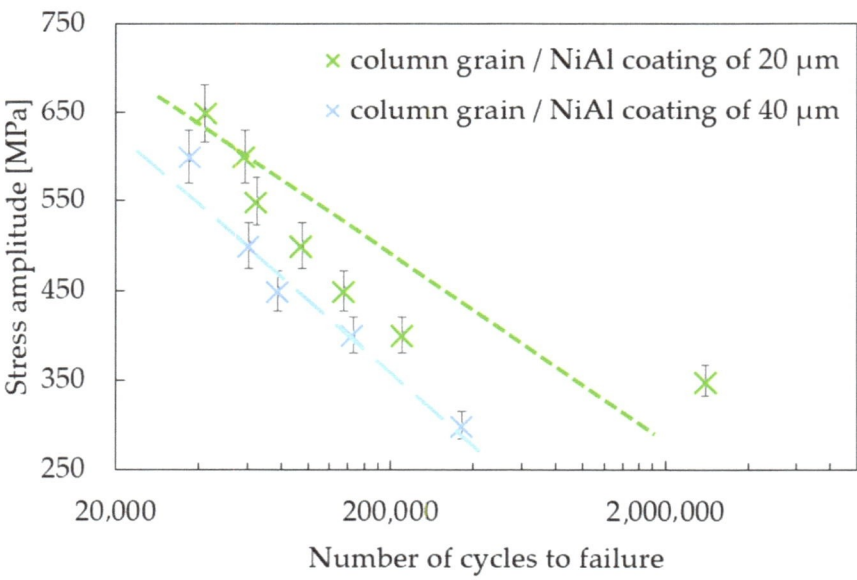

Figure 7. S-N curves for the coated, column-grained MAR 247 nickel-based superalloy.

A comparison of the fatigue response of three different microstructures and two thicknesses of the NiAl-coated MAR 247 nickel-based superalloy is presented in Figures 8 and 9. It was found that the fine-grained MAR 247 nickel-based superalloy can be characterized by the best fatigue response either for the 20 or 40 μm-thick coating. It is widely known that the microstructure refinement slightly affects the fatigue behavior of the nickel-based superalloys [36]. A similar result was reported by Mukhtarov et al. [37], where a 718 nickel-based superalloy with a different grain size (0.1–10 μm) was subjected to the low cyclic fatigue at a temperature of 23 °C. It was observed that the lifetime of the 718 nickel-based superalloy was approximately the same, regardless of the grain size. Therefore, one can conclude that the main factor affecting the fatigue performance of the MAR 247 nickel-based superalloy could be attributed to the coating itself. The results presented in Figures 8 and 9 and those reported by Akebono et al. [35] allowed us to clearly state that coating thickness strongly affected the fatigue properties, and as a consequence, when the thinner coating thickness was applied, a higher fatigue strength was achieved. Such behavior was related to the fatigue crack propagation through many defects on the coated surface in which the number of the coating defects and their sizes were mainly dependent on the coating thickness. For the thicker coatings, a larger number of defects with greater sizes were observed. Consequentially, the coated specimens with thinner coatings indicated a higher fatigue strength.

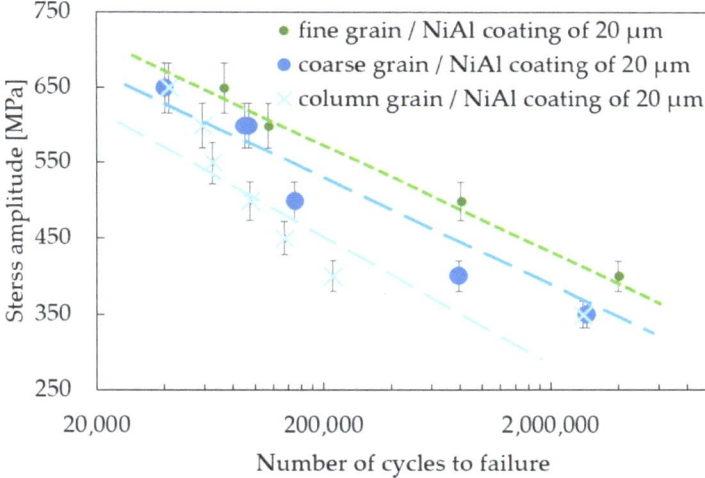

Figure 8. S-N curves for the MAR 247 nickel-based superalloy coated with 20 μm-thick NiAl.

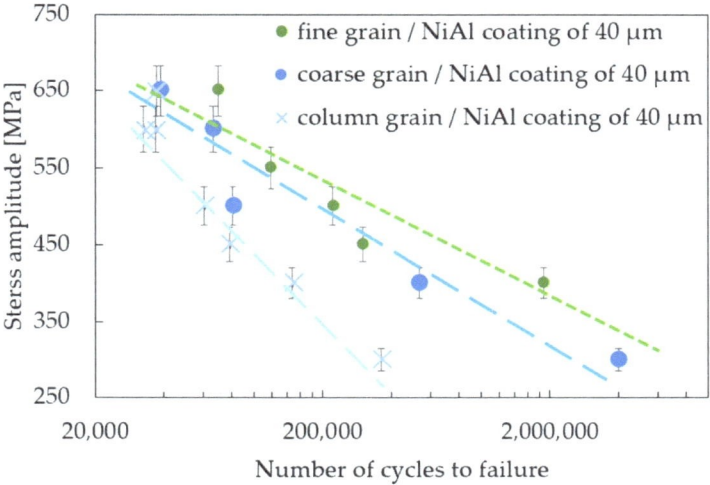

Figure 9. S-N curves for the MAR 247 nickel-based superalloy coated with 40 μm-thick NiAl.

The eddy currents measurements were performed for specimens tested in the low-cycle fatigue regime (LCF) under four cyclic stress amplitudes ranging from 450 to 650 MPa and a frequency equal to 240 Hz (Figure 10a–c). The measurements were performed in situ in six different stages of the fatigue, i.e., during fatigue tests for every 10,000 cycles and up to 60,000 cycles. When the specimen reached the specific number of cycles, the fatigue test was interrupted, and measurements were performed. After reaching 60,000 cycles, the specimen was tested to failure without subsequent EC measurements. It was found that the electrical conductivity value decreased with the number of cycles to failure, which can be attributed to the gradual degradation of the material during fatigue testing (Figure 11). Such effect was observed for all specimens, regardless of their initial microstructure or the coating thickness. The eddy current method enabled the identification of fatigue cracks at the stage preceding a development of the dominant crack in the period of 60,000 loading cycles. The significant decrease in the electrical conductivity value of about 0.5 indicates a crack initiation in the scanned area. The most important condition for crack identification is a dimension of its depth, which should exceed 50 µm. Therefore, during the in situ fatigue crack diagnostics of structural elements, the EC method was found to be useful in the effective identification of the potential areas of crack initiation (Figure 10c). Nondestructive techniques have been widely used to monitor fatigue cracks. In a comprehensive review prepared by Kong et al. [38], the eddy current method was found to be as promising and more sensitive than many other techniques (X-ray radiography, ultrasound testing, thermography, acoustic emission and digital image correlation) for potential crack localization. The eddy current method is particularly successful in fatigue damage monitoring, since it enables the determination of a number of cycles to the crack initiation [39–41]. It was shown by Potthoff et al. [42] that this method could be effectively used for the monitoring of low-cycle fatigue damage of the cast metals. In this study, the next step was taken towards extending this method to damage monitoring in coated materials.

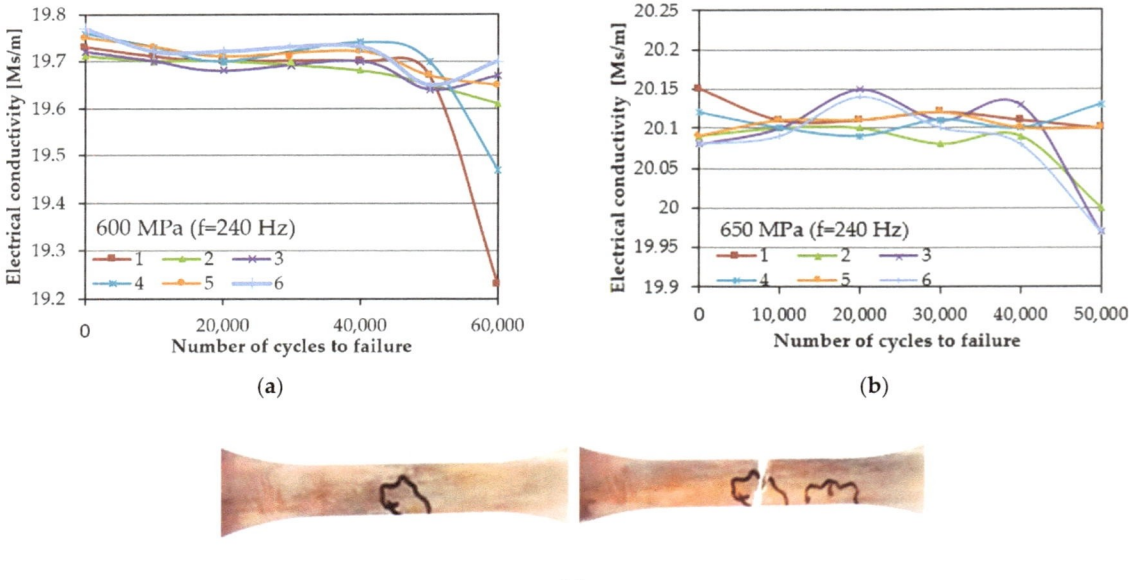

Figure 10. The exemplary results of electrical conductivity evolution measured in the selected loading cycles for two different levels of loading (**a**,**b**). Localized area of the potential crack and fracture of the specimen found after fatigue test (**c**).

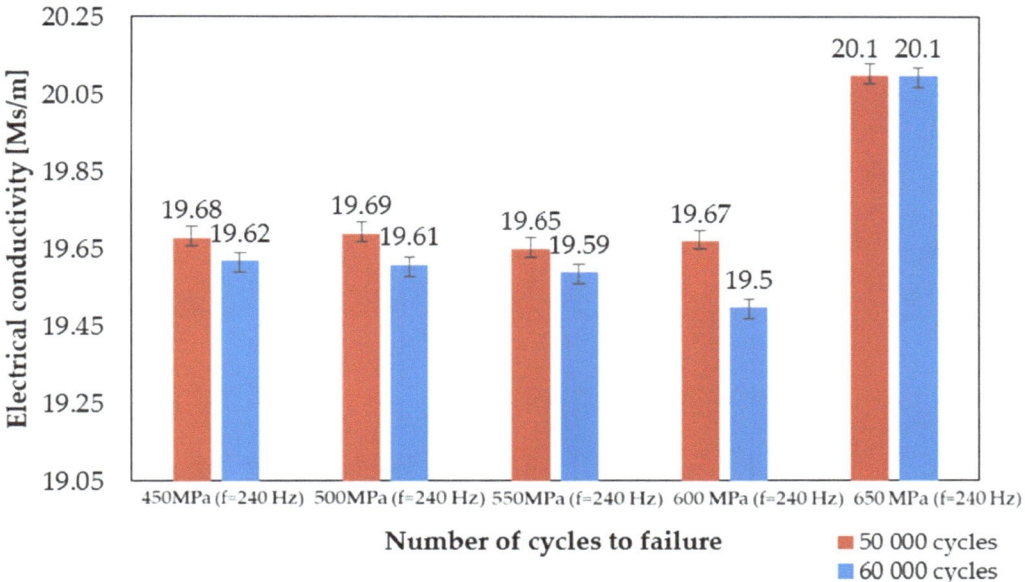

Figure 11. Comparison of electrical conductivity evolution measurements in the selected loading cycles.

The fracture surface inspections on specimens subjected to high-cycle fatigue tests were carried out. The results are presented in Figure 12. One can indicate that the CVD process parameters were successfully applied as the NiAl coating remained adhered to the substrate after decohesion within the specimen gauge length. Moreover, no particular damage or decohesion between the sublayers was observed in the area close to the dominant propagating crack for all the specimens employed. The occurrence of voids on the fracture surface was not observed either. The fine-grained fracture surface consisted of a developed structure with large and small dimples (Figure 12a,b marked with arrows). Such microstructure reflected a plastic deformation of 8% during uniaxial tensile tests and the best relative fatigue performance. The cracks found within the surface of the coarse structure (Figure 12c,d marked with arrows) were related to the brittleness of the grain boundaries. The crack formation of the nickel-based superalloys at ambient temperatures was mainly determined by the existence of carbide precipitates on the grain boundaries [42]. The occurrence of $M_{23}C_6$ carbides precipitated at the grain boundaries could further improve the mechanical properties and prevent against grain boundary sliding [43]. Column structured (Figure 12e,f marked with arrows) fracture surfaces were characterized by a striation pattern that is typical for the transgranular fracture. These striation patterns on the fracture surface were probably caused by the presence of large individual grains in the column-grained structure. It was concluded that grain boundary crack formation mainly determines the minimum ductility of the MAR 247 nickel-based superalloy. The cracks might nucleate from both the coating and the grain boundary carbides. However, the area on the coating where the initial crack occurred might be the potential place of crack propagation. The NiAl coating applied did not change the deformation mechanisms of the nickel-based superalloys, which are mainly determined by the deformation conditions and initial microstructure of the alloy investigated.

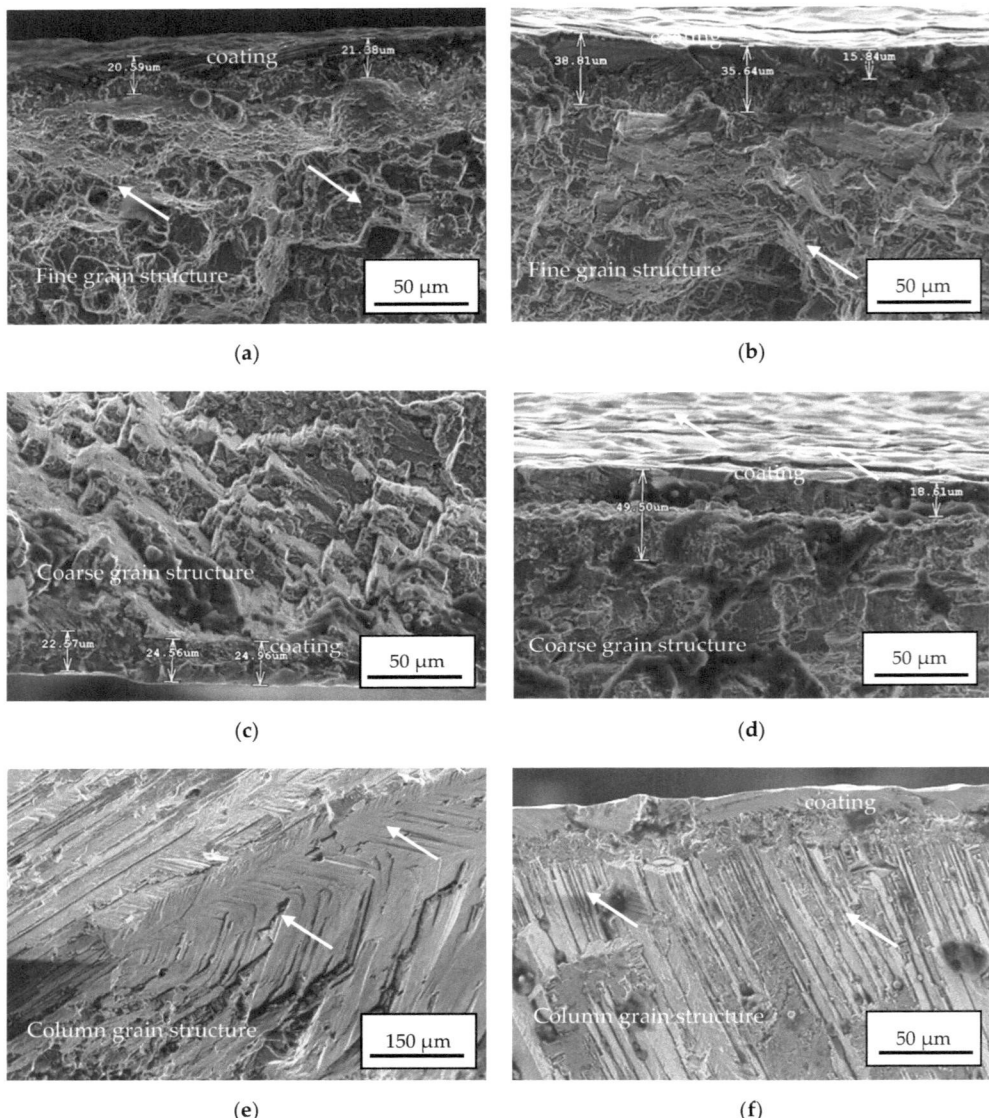

Figure 12. Fracture surfaces of the coated MAR 247 nickel-based superalloy: (**a**) fine grain/NiAl coating of 20 μm; (**b**) fine grain/NiAl coating of 40 μm; (**c**) coarse grain/NiAl coating of 20 μm; (**d**) coarse grain/NiAl coating of 40 μm; (**e**) column grain/NiAl coating of 20 μm; (**f**) column grain/NiAl coating of 40 μm observed by using scanning electron microscope.

4. Conclusions

The LCF performance of the aluminide layer-coated and as-received MAR 247 nickel superalloy with three different initial microstructures (fine grain, coarse grain and column-structured grain) was monitored using a nondestructive, eddy current method. The fatigue performance was mainly driven by the initial microstructure of the MAR 247 nickel superalloy and the thickness of the aluminide layer. The fine-grained MAR 247 nickel-based superalloy was characterized by the best fatigue response for the 20 or 40 μm-thick coating.

It was also found that the eddy current method could be successfully used to monitor the low-cycle fatigue performance of the coated nickel-based superalloys. The changes in electrical conductivity could be used as an effective indicator of damage development in the structural materials. The methodology proposed enables the in situ eddy current measurements that make it possible to successfully identify the area with potential crack initiation and its propagation within a range of cycles representing LCF.

Author Contributions: Conceptualization, M.K. and D.K.; methodology, D.K., M.K., K.W. and C.S.; formal analysis, M.K., C.S. and Z.L.K.; writing—original draft preparation M.K.; writing—review and editing, Z.L.K. All authors have read and agreed to the published version of the manuscript.

Funding: The authors gratefully acknowledge the funding by The National Centre for Research and Development, Poland, under Program for Applied Research, grant no. 178781.

Institutional Review Board Statement: Not applicable.

Informed Consent Statement: Not applicable.

Data Availability Statement: Data available in a publicly accessible repository.

Acknowledgments: The authors are grateful to Mirosław Wyszkowski, Andrzej Chojnacki and Izabela Barwińska from the Institute of Fundamental Technological Research of the Polish Academy of Sciences for their great support of the experimental work.

Conflicts of Interest: The authors declare no conflict of interest.

References

1. Agarwal, D.C.; Brill, U. High-Temperature-Strength Nickel Alloy. *Adv. Mater. Process.* **2000**, *158*, 31–34.
2. Wanhill, R.J.H. Fatigue of air supply manifold support rod in military jet engines. *J. Fail. Anal. Prev.* **2004**, *4*, 53–61. [CrossRef]
3. Kaufman, M. Properties of cast Mar-M-247 for Turbine Blisk applications. *Superalloys* **1984**, 43–52. [CrossRef]
4. Fritz, L.J.; Koster, W.P. *Tensile and Creep Rupture Properties of (1) Uncoated and (2) Coated Engineering Alloys at Elevated Temperatures*; NASA Technical Report, NAS CR-135138; NASA: Washington, DC, USA, 1977.
5. NCO. *Alloy IN-738 Technical Data*; The International Nickel Company, Inc.: New York, NY, USA. Available online: https://www.nickelinstitute.org/media/1709/in_738alloy_preliminarydata_497_.pdf (accessed on 1 December 2020).
6. Wee, S.; Do, J.; Kim, K.; Lee, C.; Seok, C.; Choi, B.-G.; Choi, Y.; Kim, W. Review on Mechanical Thermal Properties of Superalloys and Thermal Barrier Coating Used in Gas Turbines. *Appl. Sci.* **2020**, *10*, 5476. [CrossRef]
7. Garimella, L.; Liaw, P.K.; Klarstrom, D.L. Fatigue behavior in nickel-based superalloys: A literature review. *JOM* **1997**, *49*, 67. [CrossRef]
8. Kawahara, Y. Application of High Temperature Corrosion-Resistant Materials and Coatings under Severe Corrosive Environment in Waste-to-Energy Boilers. *J. Therm. Spray Technol.* **2007**, *16*, 202–213. [CrossRef]
9. Sadeghi, E.; Markocsan, N.; Joshi, S. Advances in Corrosion-Resistant Thermal Spray Coatings for Renewable Energy Power Plants: Part II—Effect of Environment and Outlook. *J. Therm. Spray Technol.* **2019**, *28*, 1789–1850. [CrossRef]
10. Jozwik, P.; Polkowski, W.; Bojar, Z. Applications of Ni3Al Based Intermetallic Alloys—Current Stage and Potential Perceptivities. *Materials* **2015**, *8*, 2537–2568. [CrossRef]
11. Cinca, N.; Lima, C.R.C.; Guilemany, J.M. An overview of intermetallics research and application: Status of thermal spray coatings. *J. Mater. Res. Technol.* **2013**, *2*, 75–86. [CrossRef]
12. Bochenek, K.; Basista, M. Advances in processing of NiAl intermetallic alloys and composites for high temperature aerospace applications. *Prog. Aerosp. Sci.* **2015**, *79*, 136–146. [CrossRef]
13. Curry, N.; Markocsan, N.; Li, X.-H.; Tricoire, A.; Dorfman, M. Next Generation Thermal Barrier Coatings for the Gas Turbine Industry. *J. Therm. Spray Technol.* **2011**, *20*, 108–115. [CrossRef]
14. Stekovic, S. Low Cycle Fatigue and Fracture of a Coated Superalloy CMSX-4. In *Fracture of Nano and Engineering Materials and Structures*; Gdoutos, E.E., Ed.; Springer: Dordrecht, The Netherlands, 2006. [CrossRef]
15. Okazaki, M. High-temperature strength of Ni-base superalloy coatings. *Sci. Technol. Adv. Mater.* **2001**, *2*, 357–366. [CrossRef]
16. Rodriguez, P.; Mannan, S.L. High temperature low cycle fatigue. *Sadhana* **1995**, *20*, 123–164. [CrossRef]
17. Šmíd, M.; Horník, V.; Hutař, P.; Hrbáček, K.; Kunz, L. High Cycle Fatigue Damage Mechanisms of MAR-M 247 Superalloy at High Temperatures. *Trans. Indian Inst. Met.* **2016**, *69*, 393–397. [CrossRef]
18. Šmíd, M.; Horník, V.; Kunz, L.; Hrbácek, K.; Hutar, P. High Cycle Fatigue Data Transferability of MAR-M247 Superalloy from Separately Cast Specimens to Real Gas Turbine Blade. *Metals* **2020**, *10*, 1460. [CrossRef]
19. Maggiani, G.; Roy, M.; Colantoni, S.; Withers, P.; Stramare, S.; Monti, C.; Bernardini, L. High temperature low cycle fatigue characterization of equiaxed MAR-M-247. *Int. J. Fatigue* **2019**, *123*, 225–237. [CrossRef]

20. Madhukar, S.; Reddy, B.R.H.; Kumar, G.A.; Naik, R.P. A Study on Improvement of Fatigue Life of materials by Surface Coatings Int. *J. Curr. Eng. Technol.* **2018**, *8*, 5–9. [CrossRef]
21. Blachnio, J.; Bogman, M. A non-destructive method to assess condition of gas turbine blades, based on the analysis of blade-surface images. *Russ. J. Nondestruct. Test.* **2010**, *46*, 860–866. [CrossRef]
22. Bogdan, M.; Błachnio, J.; Kułaszka, A.; Derlatka, M. Assessing the Condition of Gas Turbine Rotor Blades with the Optoelectronic and Thermographic Methods. *Metals* **2019**, *9*, 31. [CrossRef]
23. Vijaya Lakshmi, M.R.; Mondal, A.K.; Jadhav, C.K.; Sreedhar, S. Quantitative NDE of Aero Engine Turbine Rotor Blade—A case study. In Proceedings of the National Seminar & Exhibition on Non-Destructive Evaluation (NDE 2011), Chennai, India, 6–10 December 2011; pp. 301–304.
24. Gao, C.; Meeker, W.C.; Mayton, D. Detecting cracks in aircraft engine fan blades using vibrothermography nondestructive evaluation. *Reliab. Eng. Syst. Saf.* **2014**, *131*, 229–235. [CrossRef]
25. Gryguc, A.; Behravesh, S.; Jahed, H.; Wells, M.; Williams, B.; Gruber, R.; Duquett, A.; Sparrow, T.; Lambrou, M.; Su, X. Effect of thermomechanical processing defects on fatigue and fracture behaviour of forged magnesium. *Fratt. ed Integrità Strutt.* **2020**, *15*, 213–227. [CrossRef]
26. Kukla, D.; Kopec, M.; Kowalewski, Z.L.; Politis, D.J.; Jozwiak, S.; Senderowski, C. Thermal Barrier Stability and Wear Behavior of CVD Deposited Aluminide Coatings for MAR 247 Nickel Superalloy. *Materials* **2020**, *12*, 3863. [CrossRef]
27. Kopec, M.; Kukla, D.; Yuan, X.; Rejmer, W.; Kowalewski, Z.L.; Senderowski, C. Aluminide Thermal Barrier Coating for High Temperature Performance of MAR 247 Nickel Based Superalloy. *Coatings* **2021**, *11*, 48. [CrossRef]
28. Basak, A.; Das, S. Microstructure of nickel-base superalloy MAR-M247 additively manufactured through scanning laser epitaxy (SLE). *J. Alloys Compd.* **2017**, *705*, 806–816. [CrossRef]
29. Liu, F.; Lin, X.; Yang, G.; Song, M.; Chen, J.; Huang, W. Microstructure and residual stress of laser rapid formed Inconel 718 nickel-base superalloy. *Opt. Laser Technol.* **2011**, *43*, 208–213. [CrossRef]
30. Zhan, Z.; He, Y.; Li, L.; Liu, H.; Dai, Y. Low-temperature formation and oxidation resistance of ultrafine aluminide coatings on Ni-base superalloy. *Surf. Coat. Technol.* **2009**, *203*, 2337–2342. [CrossRef]
31. Sulak, I.; Obrtlik, K.; Celko, L. High-temperature low-cycle fatigue behaviour of HIP treated and untreated superalloy MAR-M247. *Kovove Mater.* **2016**, *54*, 471–481. [CrossRef]
32. Eder, M.A.; Haselbach, P.U.; Mishin, O.V. Effects of Coatings on the High-Cycle Fatigue Life of Threaded Steel Samples. *J. Mater. Eng. Perform.* **2018**, *27*, 3184–3198. [CrossRef]
33. Bergengren, Y.; Melander, A. An Experimental and Theoretical-Study of the Fatigue Properties of Hot-Dip-Galvanized High-Strength Sheet Steel. *Int. J. Fatigue* **1992**, *14*, 154–162. [CrossRef]
34. Akebono, H.; Komotori, J.; Shimizu, M. Effect of coating microstructure on the fatigue properties of steel thermally sprayed with Ni-based self-fluxing alloy. *Int. J. Fatigue* **2008**, *30*, 814–821. [CrossRef]
35. Akebono, H.; Komotori, J.; Suzuki, H. The effect of coating thickness on fatigue properties of steel thermally sprayed with ni-based self-fluxing alloy Int. *J. Mod. Phys. B* **2006**, *20*, 3599–3604. [CrossRef]
36. Antolovich, S.D. Microstructural aspects of fatigue in Ni-basesuperalloys. *Philos. Trans. R. Soc. A* **2015**, *373*, 20140128. [CrossRef] [PubMed]
37. Mukhtarov, S.; Utyashev, F. Effect of microstructure refinement on low cycle fatigue behavior of Alloy 718. *MATEC Web Conf.* **2014**, *14*, 04001. [CrossRef]
38. Kong, Y.; Bennett, C.J.; Hyde, C.J. A review of non-destructive testing techniques for the in-situ investigation of fretting fatigue cracks. *Mater. Des.* **2020**, *196*, 109093. [CrossRef]
39. Zilberstein, V.; Walrath, K.; Grundy, D.; Schlicker, D.; Goldfine, N.; Abramovici, E.; Yentzer, T. MWM eddy-current arrays for crack initiation and growth monitoring. *Int. J. Fatigue* **2003**, *25*, 1147–1155. [CrossRef]
40. Jiao, S.; Cheng, L.; Li, X.; Li, P.; Ding, H. Monitoring fatigue cracks of a metal structure using an eddy current sensor EURASIP. *J. Wirel. Commun. Netw.* **2016**, *188*, 1–14. [CrossRef]
41. Potthoff, M.; Peterseim, J.; Thale, W. Monitoring of Low Cycle Fatigue Damage with Eddy Current. In Proceedings of the 19th World Conference on Non-Destructive Testing, Munich, Germany, 13–17 June 2016.
42. Sajjadi, S.; Zebarjad, S.M. Study of fracture mechanisms of a Ni-Base superalloy at different temperatures. *J. Achiev. Mater. Manuf.* **2006**, *18*, 227–230.
43. Sajjadi, S.; Nategh, S.; Guthrie, R.I.L. Study of microstructure and mechanical properties of high performance Ni-base superalloy GTD-111. *Mater. Sci. Eng. A* **2002**, *325*, 484–489. [CrossRef]

Article

Influence of Anodization Temperature on Geometrical and Optical Properties of Porous Anodic Alumina(PAA)-Based Photonic Structures

Ewelina Białek [1], Maksymilian Włodarski [2] and Małgorzata Norek [1,*]

[1] Institute of Materials Science and Engineering, Faculty of Advanced Technologies and Chemistry, Military University of Technology, Str. gen Sylwestra Kaliskiego 2, 00-908 Warsaw, Poland; ewelina.bialek2@gmail.com
[2] Institute of Optoelectronics, Military University of Technology, Str. gen. Sylwestra Kaliskiego 2, 00-908 Warsaw, Poland; maksymilian.wlodarski@wat.edu.pl
* Correspondence: malgorzata.norek@wat.edu.pl

Received: 23 June 2020; Accepted: 13 July 2020; Published: 16 July 2020

Abstract: In this work, the influence of a wide range anodizing temperature (5–30 °C) on the growth and optical properties of PAA-based distributed Bragg reflector (DBR) was studied. It was demonstrated that above 10 °C both structural and photonic properties of the DBRs strongly deteriorates: the photonic stop bands (PSBs) decay, broaden, and split, which is accompanied by the red shift of the PSBs. However, at 30 °C, new bands in transmission spectra appear including one strong and symmetric peak in the mid-infrared (MIR) spectral region. The PSB in the MIR region is further improved by a small modification of the pulse sequence which smoothen and sharpen the interfaces between consecutive low and high refractive index layers. This is a first report on PAA-based DBR with a good quality PSB in MIR. Moreover, it was shown that in designing good quality DBRs a steady current recovery after subsequent application of high potential (U_H) pulses is more important than large contrast between low and high potential pulses (U_H-U_L contrast). Smaller U_H-U_L contrast helps to better control the current evolution during pulse anodization. Furthermore, the lower PSB intensity owing to the smaller U_H-U_L contrast can be partially compensated by the higher anodizing temperature.

Keywords: porous anodic alumina (PAA); pulse anodization; distributed Bragg reflector (DBR); PAA-based DBR; transmission spectra; photonic stop band (PSB); temperature

1. Introduction

Porous anodic alumina (PAA) is a multifunctional porous ceramic coating prepared by anodization of aluminum. Its geometrical parameters, such as interpore distance (D_c) and pore diameter (D_p), can be controlled by electrochemical conditions including type and concentration of electrolyte, temperature, applied voltage, and anodization time [1,2]. PAA with long-range hexagonally ordered and parallel pores is usually formed under self-ordering regimes, which are defined by narrow process windows characteristic for a given electrolyte [3,4]. Out of these regimes, the pore arrangement strongly deteriorates. The best hexagonal pore ordering is, however, obtained upon anodization conducted close to the so-called critical voltage, where high current/high electric field strength conditions are operative [5,6]. The larger the dissociation constant for a given electrolyte, the lower is the critical voltage [7]. Stability of the anodization is regulated also by temperature: the higher the temperature, the lower the critical voltage. The applied voltage determines mostly the D_c [8,9], while temperature—both D_c and D_p [10,11]. Porosity of PAA is related to both interpore distance and pore diameter via the equation: $P = \frac{\pi}{2\sqrt{3}}\left(\frac{D_p}{D_c}\right)^2$ [12]. PAA can be used as a template to fabricate various functional nanostructures [13–15] or itself is a functional material used in gas separation [16], medicine (tissue engineering) [17], or electronics (supercapacitors) [18]. One of the recent applications of PAA is related with a possibility to engineer PAA-based photonic structures.

The photonic structures based on porous materials are usually one-dimensional (1D) photonic crystals (PCs) built of many alternating low and high refractive index (RI) layers and the thickness adequate to obtain the enhancement of a selected wavelength (λ), as a result of constructive interference of waves reflected from the interface between the neighboring layers (the photonic stopband, PSB) [19,20]. The PSB is thus related to the wavelength ranges where the material demonstrates high reflectivity and low transmittance of light. The refractive index of porous layer is strictly related with its porosity (P): RI decreases as P increases [21]. Therefore, one of the important issue in designing this type of photonic crystals is a precise control over the porosity of the alternating and subsequent layers. The larger the refractive index contrast (the difference between the low and high RI of alternating layers), the more light is reflected from the layer boundary, and the more intensive PSB at a given λ can be obtained. Spectral position of PSB is very sensitive to a slight change of refractive index of a medium, and this property is a base for engineering a broad range of optical sensors [22–25].

To obtain PAA with periodically variable porosity of alternating layers, pulse anodization is applied which relies on periodic change of anodization parameters, such as voltage or current density [26,27]. The pulse sequence can be modified accordingly (different anodization modes, different current/voltage values of the generated cycles, various shape of the pulses including saw-like or pseudo-stepwise anodization waves, various ramp rate between high (U_H) and low potential (U_L), different duration of the pulses, variable number of pulses, etc.) to design a desired pore architecture and to mold strong and narrow resonances at a given spectral range. During pulse anodization other factors—such as electrolyte composition, concentration, and temperature—need to be carefully selected in order to maintain the self-ordering conditions without uncontrolled pore branching, or burning effects. PAA-based PCs developed so far were synthesized prevalently in sulfuric [28–32] and oxalic [33–36] electrolytes, using galvanostatic (current density control) or potentiostatic (voltage control) mode. First pulse anodization experiments in oxalic acid appeared to be quite challenging as compared to those in sulfuric acid owing to a relatively dense and compact barrier layer formed under high potential pulses (U_H = 55 V) [37]. The dense barrier layer prevented passage of ions under application of subsequent low potential pulses (U_L = 40 V) [37]. However, it was later shown that a uniform growth of alternating low and high RI layers is also possible upon application of a stepwise decrease of U_H to U_L, which allowed for a continuous barrier layer thinning [34]. Moreover, thanks to the application of a suitable voltage ramp rate, it was possible to increase the contrast between the U_H (50 V) and U_L (20 V) potentials, which translates into the larger refractive index contrast. Since then the influence of various pulse profiles [38,39], durations and amplitudes [40] on PAA-based PCs architecture and optical properties was studied. The optical properties of PAA-based photonic crystals were also tuned via application of mixed oxalic and phosphoric acid concentrations [41]. Mixing different concentration of phosphoric acid (0.05, 0.1, 0.2, and 0.3 M) with 0.3 M oxalic acid solution has allowed to expand the amplitudes of the applied voltage pulses (up to U_{HA} = 100 V) and to improve the PSB signal quality (smaller full width at half maximum, FWHM, and greater intensity). Despite quite extensive research on the impact of various parameters on optical properties of PAA-based PCs, little attention was paid so far to anodizing temperature, although it is a very important factor governing the growth rate of PAA layers. Furthermore, it is expected that apart from increasing the reaction speed other parameters, such as pore regularity and circularity, will be improved at higher temperatures [10]. This, in turn, should have a beneficial effect on the quality of PSBs. The photonic properties of PAA-based photonic structures were previously studied in a narrow temperature range (6–18 °C in [42] and 8–11 °C in [43]). However, it is of great interest to systematically study the impact of temperature on the properties of PAA-based PCs in a much broader range.

In this work, we analyze the impact of temperature on the quality of the PAA-based distributed Bragg reflector (DBR) structure fabricated in oxalic electrolyte in the temperature range 5–30 °C.

The impact of temperature on spectral position and intensity of PSB is systematically analyzed. It is shown that in this temperature range the photonic properties of PAA-based DBR change immensely. The PSB resonances worsen at the temperature >10 °C: the resonance peaks in transmission spectra fade away, broaden, and split, which is accompanied by the red shift of PSBs. However, at 30 °C, the peaks appear again with a strong and symmetric PSB in the mid-infrared (MIR) spectral region. Moreover, the photonic properties in the MIR are improved by a small modification of pulse sequence which sharpens the boundaries between low and high RI segments in PAA. This is a first report on production of PAA-based DBR with a good quality PSB in MIR. The results obtained in this work can extend the application of the PAA-based photonic structures up to the MIR spectral range.

2. Materials and Methods

The PAA-based distributed Bragg reflector (DBR) structures were synthesized by a pulse anodization of aluminum. High-purity aluminum foil (99.9995% Al, Puratronic, Alfa-Aesar, Haverhill, MA, USA) with a thickness of about 0.25 mm was cut into specimens (2 cm × 1 cm). Before the anodization process the Al foils were annealed under argon atmosphere at 400 °C for 2 h. Then the samples were degreased in acetone and ethanol and subsequently electropolished in a 1:4 mixture of 60% $HClO_4$ and ethanol at 0 °C, under constant voltage of 25 V, at 1 °C, and for 2.5 min. Next, the samples were rinsed with a distilled water, then ethanol, and dried. As prepared Al specimens were insulated at the back and the edges with acid resistant tape, and serve as the anode. A Pt grid was used as a cathode and the distance between both electrodes was kept constant (ca. 5 cm). A large, 1 L electrochemical cell, a powerful low-constant-temperature bath, and vigorous stirring (500 rpm) were employed in the anodizing process. Programmable DC power supply, model 62012P-600-8 Chroma, was used to control the applied voltage and the pulse parameters. The first anodization was carried out at 5 °C in 0.3 M $C_2H_2O_4$ water-based solution, at 40 V, for 20 h. As obtained alumina was chemically removed in a mixture of 6 wt % phosphoric acid and 1.8 wt % chromic acid at 60 °C for 3 h. Subsequently, pulse anodization with 20 cycles was conducted at the temperature range 5–30 °C. In general, a pulse cycle consisted of three steps: (1) a constant high voltage step (U_H = 50, 45, 40 V) applied for 360 s; (2) a gradual reduction of the voltage to 20 V (and to 30 V) at rates of 0.312, 0.234, 0.156, and 0.078 V/s (and at 0.07 V/s, respectively); (3) the anodization at a constant low voltage (U_L = 20 V) for 480 s (one sample was anodized under U_L = 30 V and the pulse duration of 3600 s). After the pulse anodization was completed, the remaining aluminum substrate was selectively removed in a saturated solution of $HCl/CuCl_2$.

Structural characterization of the PAA-based photonic structures was made using a field-emission scanning electron microscope FE-SEM (AMETEK, Inc., Mahwah, NJ, USA) equipped with energy dispersive X-ray spectrometer (EDS). The measurement of layer thickness was repeated three times at different points in the image of a given PAA sample and an average of the three measurements was taken to determine the initial and final d_H and d_L thickness and the total thickness of the PAA membrane (d_{tot}).

The transmission spectra were measured with two instruments. Shortwave end of the spectrum (250–2500 nm) was measured by Cary 5000 spectrometer with DRA-2500 integrating sphere from Agilent Technologies Inc., Santa Clara, CA, USA. Longwave end of the spectrum (2500–25,000 nm) was measured by Fourier-transform infrared (FTIR) spectrometer Alpha II from Bruker Corp., Billerica, MA, USA. The resolution of spectra was set to 1 nm in shortwave range and 2 cm^{-1} in longwave range.

3. Results and Discussion

Current density (i_a)—time (t) transients during pulse anodization (20 cycles) of aluminum at temperature (T) between 5–30 °C are shown in Figure 1a. The PAA-based DBR is formed by applying a series of potential pulses, comprising high potential pulse (U_H = 50 V and t_H = 360 s) followed by a low potential pulse (U_L = 20 V and t_L = 480 s). First three pulse profiles (U(V)) together with $i_a(t)$ curves recorded at 5 °C and 30 °C are demonstrated in Figure 1b. It can be seen that the current characteristics changes with temperature. At relatively low temperatures (5–15 °C), after application of U_H pulse, the current recovery effect is typical for conventional mild anodization (MA) processes, where current is large at the initial stage, goes to a minimum value passing through a current pike, and then increases gradually to reach a steady value. However, at higher temperature (20–30 °C) upon applying the U_H pulse, the current increases steeply for a short period of time and then decreases exponentially. The latter behavior is typical for hard anodization (HA) processes. The current recovery peak (i_a^{max}) after application of the successive U_H pulses starts to decay visibly for the samples anodized at T >10 °C. At 30 °C, the last current recovery peak is about 30% less intensive than the first one.

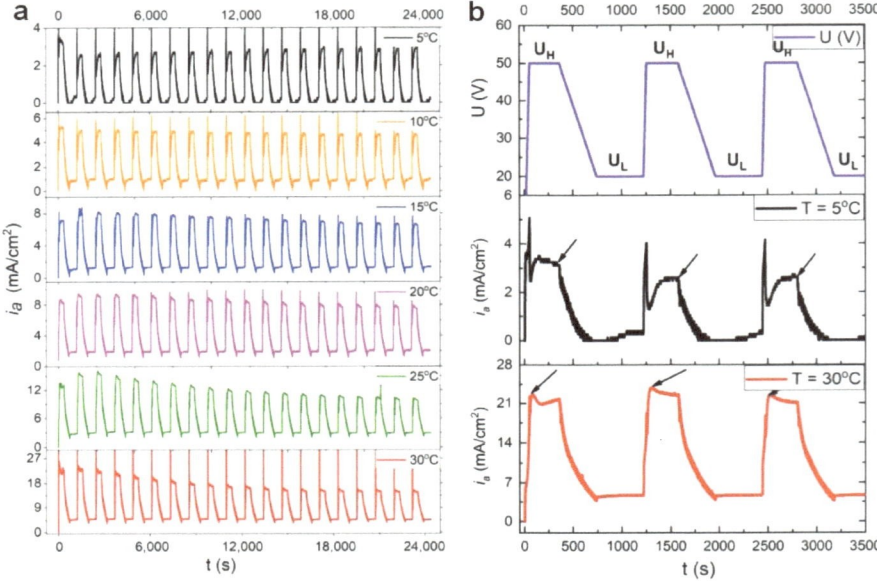

Figure 1. Current density (i_a)—time (t) transients during pulse anodization (20 cycles) of aluminum at temperatures 5–30 °C (**a**); first three U_H-U_L pulses along with the $i_a(t)$ curves for anodization at 5 °C and 30 °C (**b**). The black arrows in Figure 1b signify the i_a^{max}: depending on the type of anodization process (MA or HA) the i_a^{max} value was determined at the end or the beginning of the U_H pulse.

The unequal current recovery indicates that the total amount of charge in each U_H pulse decreases as the number of pulses increases. Since the thickness of anodic alumina is directly proportional to the net amount of charge involved in anodization reaction, it is thus expected that anodic alumina formed under the present conditions will contain segments with non-uniform thickness. Thickness of three initial and final segments (corresponding with the first and last potential pulses in the 20-cycle anodization process) formed under U_H and U_L potentials (d_H and d_L, respectively) for PAA produced at the two extreme temperatures is presented in Figure 2. As can be seen, the difference between initial and final d_H and d_L segments is much larger in the PAA-based DBR formed at 30 °C as compared to that formed at 5 °C.

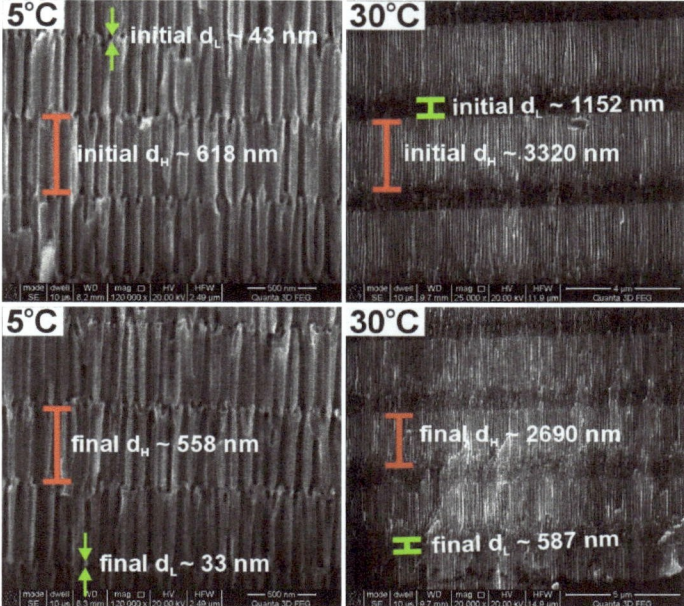

Figure 2. SEM images of a cross sectional view of the initial and final PAA segments obtained during the pulse anodization at 5 °C (the sample PAA—5 °C) and 30 °C (the sample PAA—30 °C).

The evolution of i_a^{max} (indicated by black arrows in Figure 1b) after application of the following 20 U_H pulses for all samples anodized at temperatures 5–30 °C is well visible in Figure 3a. At 5 °C and 10 °C, the i_a^{max} remains more or less at stable values during each cycle of anodizing, however, starting already from 15 °C the i_a^{max} gets weakened noticeably as the number of U_H pulses increase. At 25 °C and 30 °C, the i_a^{max} intensity drop is very significant. The behavior is followed by the increasing difference between initial and final d_H and d_L layers (Figure 3b). First of all, an increase of temperature results in thicker d_H and d_L slabs owing to the enhanced electrochemical reaction rate. The d_L difference is rather negligible in the temperature range 5–25 °C but starts to be pronounced at 30 °C. However, the difference between first and last d_H segments begins to grow substantially already at T >10 °C. The effect indicates the appearance of diffusional problems at the higher temperatures related with the extended diffusion path along the nanopores and consequently slower mass transport (the oxygen-containing anionic species such as O^{2-}, OH^-) from the electrolyte to the pore bottom [44–46]. Nevertheless, the total thickness (d_{tot}) of PAA dependence on temperature (in the 5–25 °C range) is almost linear (Figure 3c) what suggests that other, rate-limiting processes have to be also accounted for this behavior. The d_{tot} of PAA prepared at 30 °C is an exception here. The sudden collapse of the linear relationship for this sample could indicate the existence of a boundary thickness (around 54 µm), above which Al_2O_3 stops to grow.

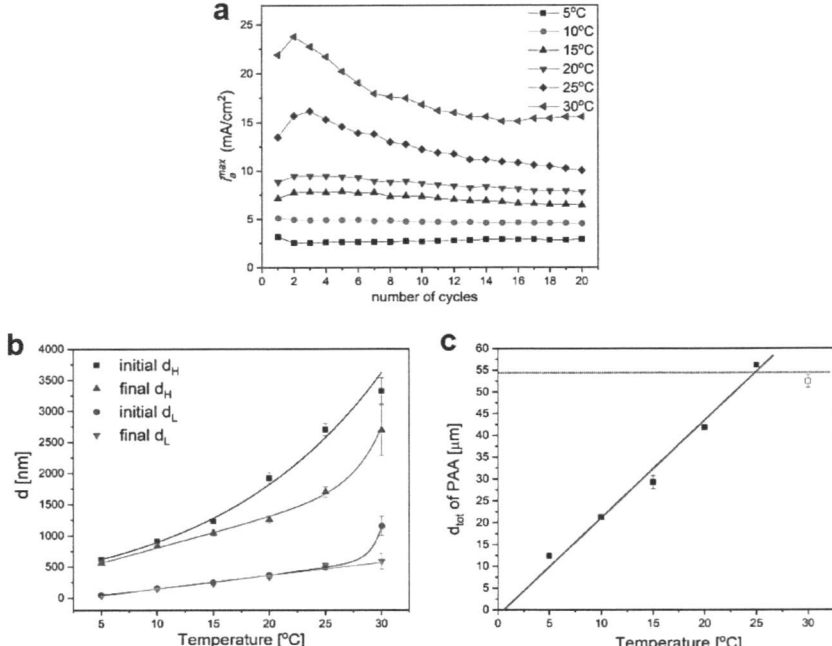

Figure 3. $i_a{}^{max}$ as a function of the number of cycles (**a**) thickness of initial and final d_H and d_L segments as a function of anodizing temperature (**b**) total thickness (d_{tot}) of PAA membranes as a function of anodization temperature (**c**).

To get deeper insight into this issue, additional PAA-based DBR structure under the following anodizing conditions was prepared: U_H = 50 V with t_H = 360 s, and U_L = 30 V with t_L = 3600 s. In order to avoid too extensive PAA growth, the process was conducted at 10 °C (therefore the sample will be further named as #PAA—10 °C). The $i_a(t)$ curves recorded during anodization of #PAA—10 °C and PAA—30 °C samples are compared in Figure 4a. Despite prolonged anodization time of the #PAA—10 °C sample (the whole process lasted ca. 24 h), the $i_a{}^{max}$ remains at quite stable values after application of subsequent U_H pulses (the recovery of current after application of the U_L pulses is very regular), in contrast to the $i_a{}^{max}$ recorded during pulse anodization of the sample PAA—30 °C. As an effect of the steady current evolution, the thickness of initial and final d_H and d_L segments in the sample #PAA—10 °C does not differ so much as in the sample PAA—30 °C (Figure 4b,c). Particularly, the thickness of initial d_L layers is larger only of about 7% from that of the final ones. Closer examination of the sample PAA—30 °C revealed that it is built only out of 15 d_L and d_H layers instead of 20, which explains its lower d_{tot} (~52 μm) than expected (Figure 4d). The resulted d_{tot} of the #PAA—10 °C is ca. 9 μm thicker (d_{tot}~61 μm) than the PAA—30 °C (Figure 4e). Moreover, despite the larger thickness of the membrane, the #PAA—10 °C photonic crystal consists of full 20 d_L and d_H layers (Figure 4e). It is therefore evident that the growth of the remaining 5 d_H and d_L layers in the PAA—30 °C sample was not stopped solely by diffusional problems related with too thick membrane. Moreover, successful formation of the 15 subsequent t_H and t_L layers in PAA—30 °C DBR indicates that the mass transport (movement of ionic species, such as O^{2-}, OH^-, Al^{3+} through the barrier layer) was not hindered by a too thick barrier layer formed at the high temperature. On the contrary, the high temperature provided sufficient driving force to overcome this barrier. Most probably a high reaction rate, which requires a continuous and relatively fast delivery of the anionic species from the bulk reservoir to the pore basis (from the electrolyte to the oxide/metal interface), was mainly responsible for reaction termination during pulse anodization at 30 °C. The electrochemical formation of PAA is determined by both diffusion-controlled and rate-controlled processes. If the PAA thickness is increased too much, the delivery is substantially delayed, and the electrochemical reaction at 30 °C is terminated due to the high reaction speed. In the case of the sample #PAA—10 °C the reaction rate is slowed down so much that the larger distances that ions have to overcome along the increasing thickness of PAA does not constitute an obstacle in the oxide formation.

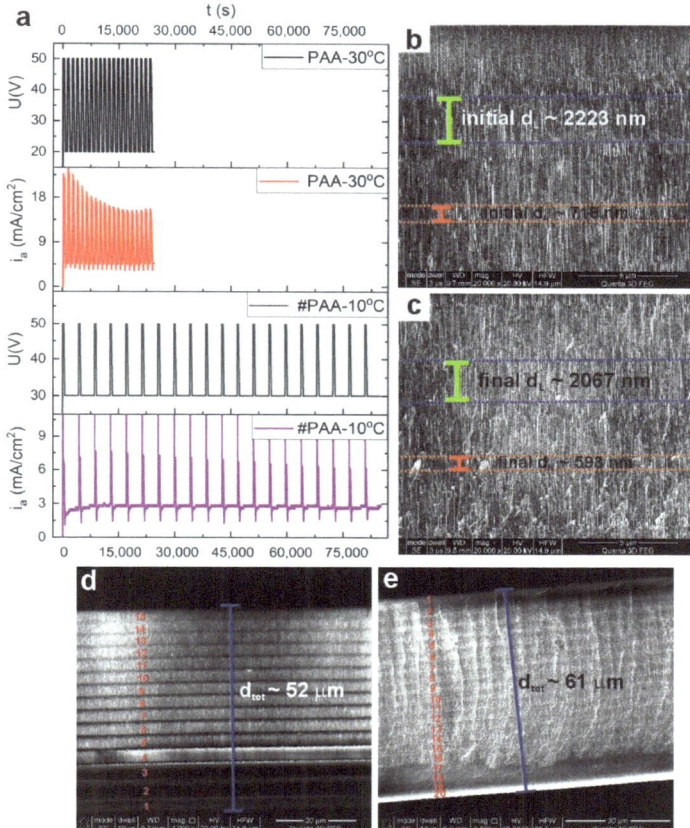

Figure 4. $i_a(t)$ curves recorded during pulse anodization (20 cycles) of the sample PAA—30 °C (U_H = 50 V with t_H = 360 s, U_L = 20 V with t_L = 480 s. at 30 °C) and #PAA—10 °C (U_H = 50 V with t_H = 360 s, U_L = 30 V with t_L = 3600 s. at 10 °C) (**a**); SEM images of initial (**b**) and final (**c**) d_H and d_L segments in the sample #PAA—10 °C; SEM image of a cross-sectional view of the whole PAA—30 °C (**d**) and #PAA—10 °C (**e**) membranes.

In Figure 5, transmission spectra of the PAA-based DBR structures fabricated at the temperature range 5–30 °C are presented. The position of photonic stop bands (PSBs) is usually determined based on Bragg–Snell law [47]

$$m\lambda = 2d\sqrt{n_{eff}^2 - n_{air}^2 sin^2\theta} \qquad (1)$$

where λ is the wavelength of a stop band, m is the order of the PSB, d is the layer thickness (periodicity), θ is the angle of incidence, n_{eff} is the effective refractive index, and n_{air} is the refractive index of air.

In the transmission spectra (T(λ)) several resonance peaks are visible which can be assigned to different orders of a given stop band (λ_i, i = 1–4, correspond to 1–4 orders of PSB; the bands were assigned to the λ_i based on the Bragg–Snell equation, assuming $\theta \sim 0$). At temperatures 5 °C and 10 °C, the peaks are distinct and narrow. As compared to the PSBs in the sample PAA—5 °C, the peaks in the spectrum of the sample PAA—10 °C are more intensive and red-shifted. Upon increasing temperature, the peaks shift further towards the red part of the spectrum, split, and become progressively broadened. In the spectrum of the PAA—25 °C they almost disappear. However, in the spectrum of the sample PAA—30 °C the peaks start to show up again: several low-order ones with lower intensity in the range 1000–2500 nm, and one located in the mid-infrared region, centered at ~4386 nm with T~0.20 (according to a commonly used subdivision scheme MIR region falls into 3–8 µm [48]). The stop band in the MIR region (assigned to λ_2 of the PSB) is very symmetric, what usually indicates a good quality photonic crystal structure. Moreover, based on the Bragg–Snell equation (for $\theta \sim 0$) and taking into account the measured periodicity ($d = d_H + d_L$) and λ_i, the n_{eff} of the studied PAA-based DBRs can be roughly estimated to be within the range of 1.13–1.57.

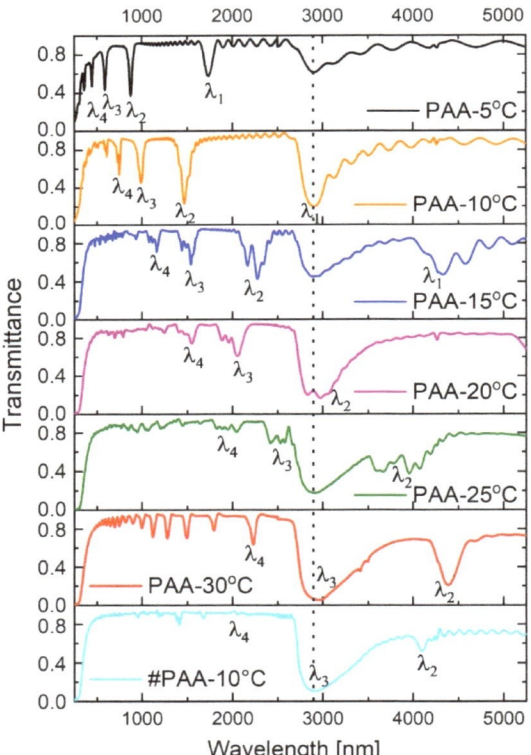

Figure 5. Transmittance spectra of PAA-based photonic structures anodized at temperature range 5–30 °C (the broad peak at around 3000 nm, present in all spectra and marked by vertical, black, dotted line, originate from OH group vibrations of adsorbed water [49]).

The spectrum of the # PAA—10 °C DBR is quite similar to the one of the sample PAA—30 °C in terms that it also shows several bands in the range 1000–2500 nm, and the one in the MIR region. The peaks are, however, much less intensive (the λ_2 band is located at around 4100 nm with T~0.5).

The PAA—30 °C photonic crystal apparently has much better optical properties than the #PAA—10 °C crystal, despite its lower number of d_L and d_H layers and larger difference in the initial and final thickness of the d_L and d_H layers. This phenomena can be due, however, to a larger refractive index contrast (Δn_{eff}) between the subsequent d_L and d_H segments. It was shown before that a stopband enlarges and sharpens when the Δn_{eff} increases [20]. The Δn_{eff} in PAA photonic material is directly related with porosity contrast (ΔP) between the d_L and d_H layers [21], and the porosity, in turn, is determined by both anodizing voltage and temperature [10,11]: the larger the applied voltage and the higher the temperature the greater is the porosity. Therefore, the ΔP in the sample PAA—30 °C is tuned and increased by both larger U_H-U_L contrast (30 V) and higher anodizing temperature (T = 30 °C) as compared to the lower U_H-U_L contrast (20 V) and lower temperature (T = 10 °C) applied to fabricate the sample #PAA—10 °C.

Morphology analysis of the PAA—30 °C DBR crystal suggests that its optical properties can be still improved by a better tailoring the d_H and d_L interfaces. In Figure 6, it can be noticed that the edge of the d_H segment that corresponds with the gradual decrease of voltage from U_H to U_L ($U_H \twoheadrightarrow U_L$ edge) is very blurred as compared to the opposite edge that corresponds with a direct voltage change from U_L to U_H ($U_L \twoheadrightarrow U_H$ edge). The clear and sharp interfaces between constitutive layers are known to be critical for highly reflective DBRs [50]. Therefore, the pulse sequence were modified accordingly by acceleration of the voltage drop rate in order to sharpen the $U_H \twoheadrightarrow U_L$ edge. In Figure 7, $i_a(t)$ curves recorded during pulse anodization (U_H = 50 V with t_H = 360 s, U_L = 20 V with t_L = 480 s, 20 cycles, T = 30 °C) with increasing $U_H \twoheadrightarrow U_L$ drop rate from 0.078 V/s up to 0.312 V/s are shown, along with the corresponding transmission spectra (the samples: PAA—30 °C_0.078, PAA—30 °C_0.156, PAA—30 °C_0.234, and PAA—30 °C_0.312, respectively).

It can be seen that upon decreasing the $U_H \twoheadrightarrow U_L$ rate from 0.0718 V/s to 0.234 V/s the intensity of the corresponding transmission dips (λ_2 and λ_3) increases. The intensity of the λ_2 band in the PAA—30 °C_0.078 sample increases from 0.20 to 0.10 for the sample PAA—30 °C_0.156 and to 0.08 for the sample PAA—30 °C_0.234. At the same time, the λ_2 shifts from 4386 nm to 3587 nm and 3602 nm in the samples anodized with the lower $U_H \twoheadrightarrow U_L$ rate (0.156 V/s and 0.234 V/s, respectively). Furthermore, the decrease of the $U_H \twoheadrightarrow U_L$ rate to 0.312 V/s deteriorates the optical properties of the PAA—30 °C_0.312 DBR: the resonance peaks split and become hardly distinguishable.

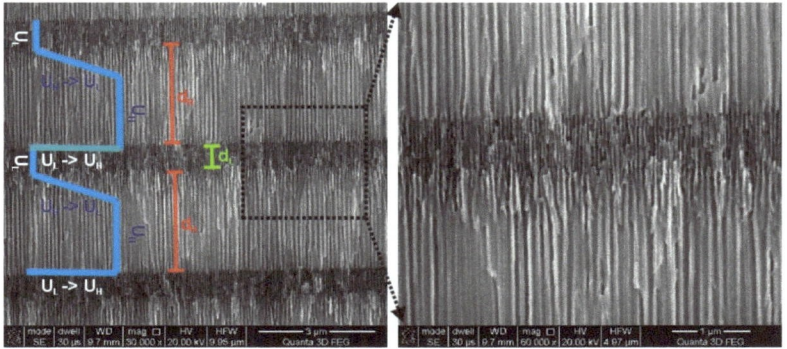

Figure 6. The $U_H \twoheadrightarrow U_L$ and $U_L \twoheadrightarrow U_H$ interfaces in the PAA—30 °C DBR.

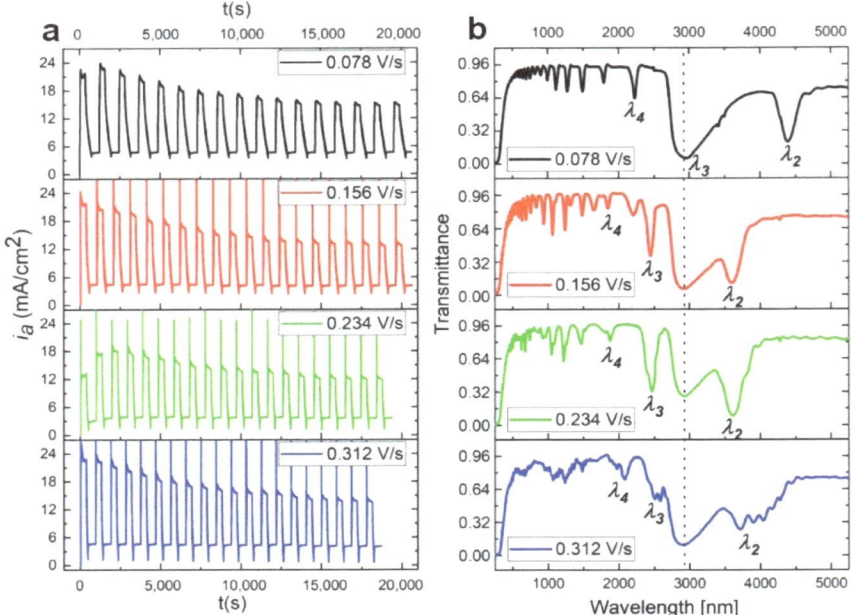

Figure 7. $i_a(t)$ curves recorded during pulse anodization (U_H = 50 V with t_H = 360 s, U_L = 20 V with t_L = 480 s, 20 cycles, at 30 °C) with decreasing $U_H \gg U_L$ drop rate form 0.078 V/s down to 0.312 V/s (**a**), the corresponding transmission spectra (the broad peak at around 3000 nm present in all spectra and marked by vertical, black, dotted line, originate from OH group vibrations of adsorbed water [49]) (**b**).

The shift of the λ_i to shorter wavelength is presumably caused by the reduced thickness of both d_H and d_L layers formed in the PAA—30 °C DBR crystals prepared under the slower $U_H \gg U_L$ rates (Figure 8a). In Figure 8b–d, SEM images of cross-sectional views of the whole PAA membranes, prepared under various $U_H \gg U_L$ rates, are shown. It can be observed that the number of d_H and d_L segments increases as the $U_H \gg U_L$ rate decreases (Figure 8b–d). In the sample PAA—30 °C_0.312 almost all segments (19 out of 20) were formed. The d_{tot} varies also with the $U_H \gg U_L$ rate, but does not exceed 54 µm for the PAA—30 °C_0.312 sample. It seems thus that the 54 µm is indeed a limit thickness for PAA-based DBR grown at 30 °C. On the other hand, it can be expected that lowering T by few degrees (between 26 °C and 29 °C) will help to prepare DBR built out of full 20 d_H and d_L segments. This, in turn, will contribute to optimization of the photonic characteristics of PAA-based DBR with PSB resonances in MIR. Summarizing: the better optical quality of the samples PAA—30 °C_0.156 and PAA—30 °C_0.234 DBRs as compared to that of the PAA—30 °C_0.078 sample is caused by a larger number of constitutive d_H and d_L layers and by much sharper interfaces between d_H and d_L layers from the $U_H \gg U_L$ side (Figure 8e–h). Apparently, the transmission spectra of the PAA—30 °C_0.312 DBR deteriorates owing to the large difference between initial and final d_H thickness (Figure 8a), which eliminates potential benefits emerging from the largest number of subsequent d_H and d_L segments among all PAA-based DBRs fabricated at 30 °C.

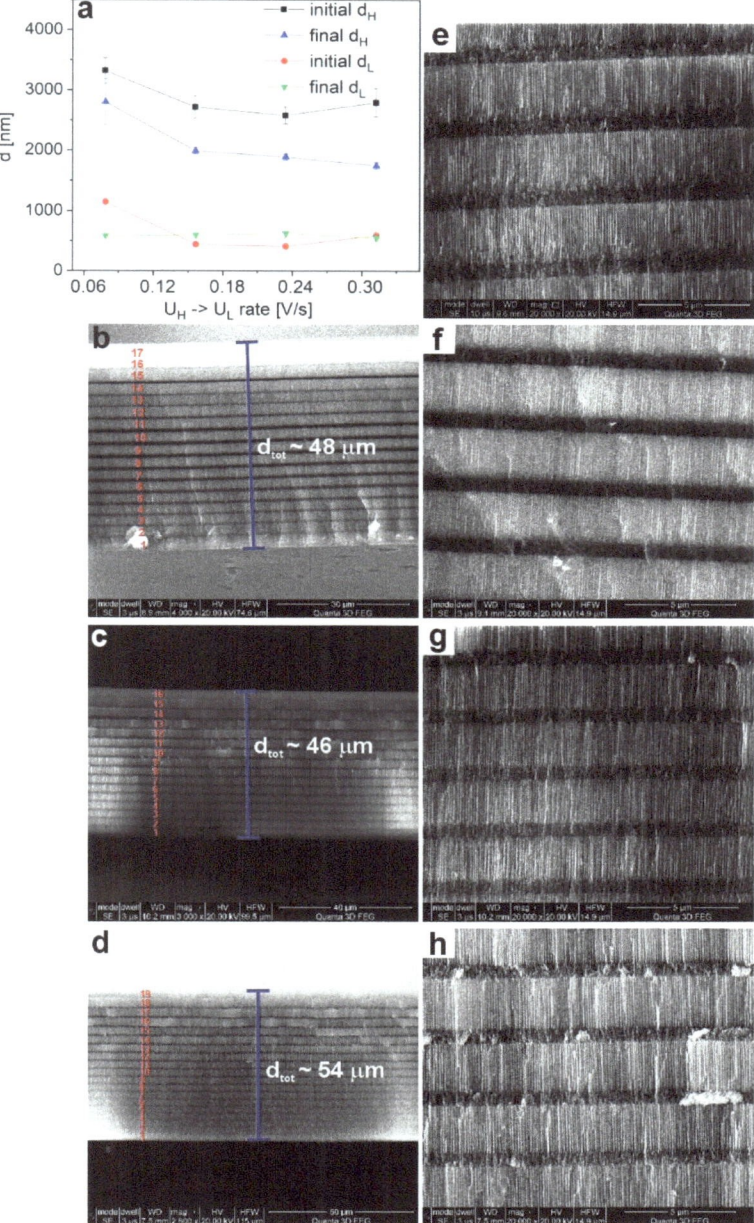

Figure 8. Thickness of d_H and d_L segments as a function of the $U_H \gg U_L$ drop rate (**a**); SEM images of a cross-sectional view of the whole PAA—30 °C_0.156 (**b**), PAA—30 °C_0.234 (**c**), and PAA—30 °C_0.312 (**d**) membrane; SEM images of middle d_H and d_L layers of the PAA—30 °C DBRs with $U_H \gg U_L$ rate of 0.078 V/s (**e**), 0.156 V/s (**f**), 0.234 V/s (**g**), and 0.312 V/s (**h**).

Based on transmission spectra in Figure 2, it can be stated that the strongest PSBs were generated in the PAA-based DBR fabricated at 10 °C. Therefore this temperature was selected to study the influence of U_H and U_L voltage on the optical properties of the PCs. In Figure 9a, the $i_a(t)$ curves for the DBRs synthesized under different values of U_H (50 V, 45 V, 40 V) and U_L (20 V, 30 V) are demonstrated, with other conditions kept as previously (the samples: PAA—10 °C_50–20, PAA—10 °C_45–20, PAA—10 °C_40–20, and PAA—10 °C_50–30, respectively). It can be observed that whereas the i_a^{max} decreases slowly with pulse cycles during anodization of the PAA—10 °C_50–20 and PAA—10 °C_45–20 DBRs, in the anodization of the PAA—10 °C_40–20 sample the i_a^{max} remains perfectly stable during all 20 cycles. The largest i_a^{max} drop is recorded for the PAA—10 °C_50–30 sample. The current evolution is reflected in the difference between initial and final thickness of d_H and d_L segments: for all PAA-based DBRs the difference is discernable, whereas in the DBR prepared under the 40–20 V all d_H and d_L layers are identical (Figure 9b). Particularly interesting is the comparison between the PAA—10 °C_40–20 and PAA—10 °C_50–30 samples. Despite the same U_H-U_L contrast (20 V), the current behavior and the resulting DBR are quite different.

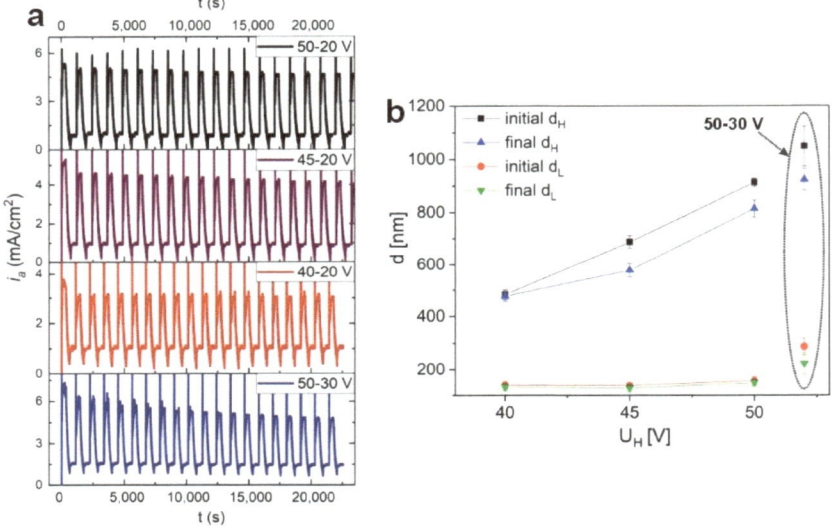

Figure 9. Current density $i_a(t)$ recorded during pulse anodization of aluminum under the following U_H-U_L potentials: 50–20 V, 45–20 V, 40–20 V, and 50–30 V (t_H = 360 s, t_L = 480 s, 20 cycles, 10 °C) (**a**); initial and final d_H and d_L thickness as a function of U_H for the U_L = 20 V (the d_H and d_L thickness determined for the PAA—10 °C_50–30 sample is placed in the graph in a dotted ellipse) (**b**).

The transmission spectrum of the PAA—10 °C sample is compared with $T(\lambda)$ spectra of the other DBRs prepared at 10 °C in Figure 10, in the 250–2500 nm spectral range. First of all, upon decreasing U_H the PSBs are shifted towards blue part of the spectrum, mostly as an effect of decreasing the d_H layer thickness (Figure 9b). The intensity of the resonance peaks decreases progressively, however, the PSBs in the samples PAA—10 °C_45–20 and PAA—10 °C_40–20 become more symmetric and narrower as compared to the PSB in the sample PAA—10 °C_50–20. The narrower and more symmetric peaks, in turn, indicate better quality of the DBR crystals. In the $T(\lambda)$ spectra of the PAA—10 °C_50–30 sample the PSBs have gone, meaning that basically no DBR structure was formed under this condition. Larger intensity of the resonance peaks in the $T(\lambda)$ spectrum of the PAA—10 °C_50–20 DBR can be associated with a larger U_H-U_L contrast (30 V) and consequently larger ΔP. However, owing to the larger U_H-U_L contrast, the current recovery (i_a^{max}) after application of the following U_H pulses gets gradually weakened, and consequently, the formed d_H and d_L segments are not perfectly uniform throughout the whole PAA membrane. This, in turn, broadens the transmission peaks. As the U_H-U_L contrast decreases (due to U_H decrease), the i_a^{max} becomes equalized (in the sample PAA—10 °C_40–20 the i_a^{max} is perfectly even) and the structural and optical properties of DBR are improved: the transmission peaks become narrower and more symmetric. However, the lower U_H-U_L contrast makes the PSB peaks less intensive owing to the smaller ΔP.

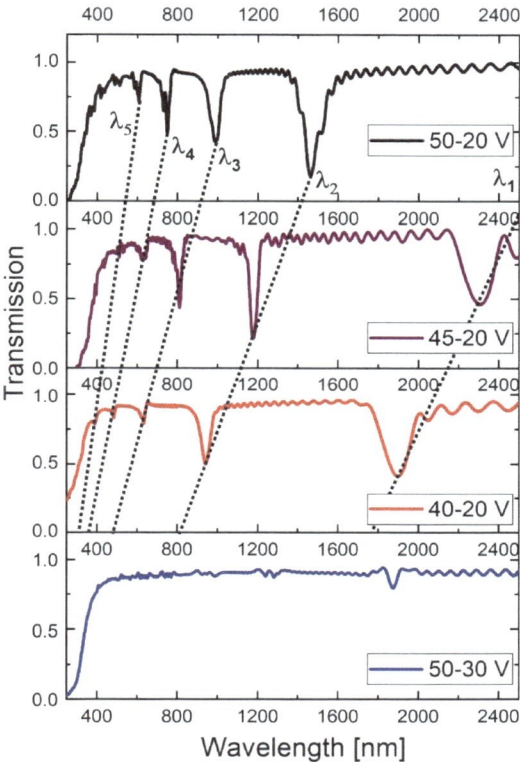

Figure 10. Transmittance spectra of PAA-based DBR fabricated at 10 °C under various U_H and U_L values.

Summarizing, when the U_H is lowered from 50 V to 40 V (while keeping the U_L constant) the quality of the crystals remains quite stable (or even improves). However, when the U_L values is increased from 20 V to 30 V (while keeping the U_H constant) the crystal quality is lost. The latter is caused by both small ΔP and the large i_a^{max} drop upon subsequent application of the U_H pulses. Based on these data, one can risk the statement that a contrast between high and low potential pulses (and thus ΔP) is less important than the steady current recovery. This statement can be also partially supported by the observation that the sample prepared under a larger U_H-U_L contrast but lower anodizing temperature (the PAA—5 °C sample) is characterized by a comparable optical quality than the one fabricated under the lower U_H-U_L contrast (the sample anodized under 40–20 V) but at higher anodizing temperature. In the latter case, however, most likely the higher temperature is the factor that improves structural (better regularity and circularity of pores) and thus optical properties PAA-based DBR.

4. Conclusions

In this work, the influence of anodization temperature from 5 °C to 30 °C range on the growth and photonic properties of PAA-based DBRs was studied. Transmission spectra were recorded to determine the position and a shape of PSBs. It was observed that above 10 °C and up to 25 °C the PSB properties strongly deteriorate as manifested in a progressive decrease, widening, and splitting of transmission peaks. However, at 30 °C the PSBs appeared again with several narrow, low intensity peaks in the 1000–2500 nm range and a strong, symmetric resonance peak in the MIR region. The photonic properties of the PAA—30 °C DBR were further improved by a small modification of the pulse sequence which sharpened the interface between d_H and d_L segments. Moreover, it was shown that larger U_H-U_L contrast helps to increase the intensity of respective PSBs owing to the larger porosity contrast (ΔP) between consecutive low and high RI layers. However, the larger U_H-U_L contrast slowed down the current recovery (i_a^{max}) after application of subsequent U_H pulses. Generally, U_H-U_L contrast seems to be less important than a steady i_a^{max} during application of consecutive pulses. Structural and optical properties of DBR anodized under 40–20 V at 10 °C were much better than the properties of the DBR synthesized under 50–30 V, despite the same U_H-U_L contrast (20 V). On the other hand, the properties were comparable with that of the DBR synthesized under 50–20 V (U_H-U_L contrast = 30 V), but at lower temperature of 5 °C. This means also that the temperature is an important factor in tailoring good quality PCs. Furthermore, the analysis performed in this work revealed that anodization at high temperature provides new conditions for designing and tailoring the PAA-based photonic structures with good photonic properties in MIR region. In fact, this was a first time the PAA-based DBR structure with a good quality PSB (relatively narrow and symmetric peak) in the MIR spectral range was fabricated. The new conditions provided by the high-temperature-pulse-anodization needs further optimization and mastering the electrochemical process (e.g., process conducted at T = 26–29 °C, various duration of U_H and U_L pulses, various U_H-U_L contrast) in order to produce PAA-based photonic structure with excellent photonic properties (strong and narrow PSB resonances) in the MIR spectral range. This work is in progress.

Author Contributions: Conceptualization, M.N.; Electrochemical synthesis of PAA-based PCs, E.B.; Transmission measurements, M.W.; SEM analysis, M.N.; Writing—original draft preparation, M.N.; All authors have read and agreed to the published version of the manuscript.

Funding: The research was financed by National Science Centre, Poland (UMO-2019/35/B/ST5/01025). The work was also supported by the statutory research funds of the Department of Functional Materials and Hydrogen Technology, Military University of Technology, Warsaw, Poland. The UV–vis–NIR spectrometers used in these studies were obtained with funds from the Polish Ministry of Science and Higher Education grant for investment in large research infrastructure no. 7044/IA/SP/2019.

Conflicts of Interest: The authors declare no conflict of interest.

References

1. Jani, A.M.; Losic, D.; Voelcker, N.H. Nanoporous anodic aluminium oxide: Advances in surface engineering and emerging applications. *Progr. Mater. Sci.* **2013**, *58*, 636–704. [CrossRef]
2. Lee, W.; Park, S.-J. Porous anodic aluminium oxide: Anodization and templated synthesis of functional nanostructures. *Chem. Rev.* **2014**, *114*, 7487–7556. [CrossRef] [PubMed]
3. Pashchanka, M.; Schneider, J.J. Self-ordering regimes of porous anodic alumina layers formed in highly diluted sulfuric acid electrolytes. *J. Phys. Chem. C* **2016**, *120*, 14590–14596. [CrossRef]
4. Li, F.; Zhang, L.; Metzger, R.M. On the growth of highly ordered pores in anodized aluminium oxide. *Chem. Mater.* **1998**, *10*, 2470–2480. [CrossRef]
5. Ono, S.; Saito, M.; Ishiguro, M.; Asoh, H. Controlling factor of self-ordering of anodic porous alumina. *J. Electrochem. Soc.* **2004**, *151*, B473–B478. [CrossRef]
6. Ono, S.; Saito, M.; Asoh, H. Self-ordering of anodic porous alumina induced by local current concentration: Burning. *Electrochem. Solid State Lett.* **2004**, *7*, B21–B24. [CrossRef]

7. Qin, X.; Zhang, J.; Meng, X.; Deng, C.; Zhang, L.; Ding, G.; Zeng, H.; Xu, X. Preparation and analysis of anodic aluminum oxide films with continuously tunable interpore distances. *Appl. Surf. Sci.* **2015**, *328*, 459–465. [CrossRef]
8. Akiya, S.; Kikuchi, T.; Natsui, S.; Sakaguchi, N.; Suzuki, R.O. Self-ordered porous alumina fabricated via phosphonic acid anodizing. *Electrochim. Acta* **2016**, *190*, 471–479. [CrossRef]
9. Kikuchi, T.; Nishinaga, O.; Natsui, S.; Suzuki, R.O. Fabrication of self-ordered porous alumina via etidronic acid anodizing and structural color generation from submicrometer-scale dimple array. *Electrochim. Acta* **2015**, *156*, 235–243. [CrossRef]
10. Zaraska, L.; Stępniowski, W.J.; Ciepiela, E.; Sulka, G.D. The effect of anodizing temperature on structural features and hexagonal arrangement of nanopores in alumina synthesized by two-step anodizing in oxalic acid. *Thin Solid Films* **2013**, *534*, 155–161. [CrossRef]
11. Stępniowski, W.J.; Bojar, Z. Synthesis of anodic aluminum oxide (AAO) at relatively high temperatures. Study of the influence of anodization conditions on the alumina structural features. *Surf. Coat. Technol.* **2011**, *206*, 265–272. [CrossRef]
12. Nielsch, K.; Choi, J.; Schwirn, K.; Wehrspohn, R.B.; Gosele, U. Self-ordering regimes of porous alumina: The 10% porosity rule. *Nano Lett.* **2002**, *2*, 677–680. [CrossRef]
13. Ikeda, H.; Iwai, M.; Nakajima, D.; Kikuchi, T.; Natsui, S.; Sakaguchi, N.; Suzuki, R.O. Nanostructural characterization of ordered gold particle arrays fabricated via aluminum anodizing, sputter coating, and dewetting. *Appl. Surf. Sci.* **2019**, *465*, 747–753. [CrossRef]
14. Norek, M.; Putkonen, M.; Zaleszczyk, W.; Budner, B.; Bojar, Z. Morphological, structural, and optical characterization of SnO2 nanotube arrays fabricated using anodic alumina (AAO) template-assisted atomic layer deposition. *Mater. Charact.* **2018**, *136*, 52–59. [CrossRef]
15. Norek, M.; Zaleszczyk, W.; Łuka, G.; Budner, B.; Zasada, D. Tailoring UV emission from a regular array of ZnO nanotubes by the geometrical parameters of the array and Al2O3 coating. *Ceram. Intern.* **2017**, *43*, 5693–5701. [CrossRef]
16. Chernova, E.; Petukhov, D.; Boytsova, O.; Alentiev, A.; Budd, B.; Yampolskii, Y.; Eliseev, A. Enhanced gas separation factors of microporous polymer constrained in the channels of anodic alumina membranes. *Sci. Rep.* **2016**, *6*, 31183. [CrossRef]
17. Poinern, E.; Ali, N.; Fawcett, D. Progress in nano-engineered anodic aluminium oxide membrane development. *Materials* **2011**, *4*, 487–526. [CrossRef] [PubMed]
18. Fernández-Menéndez, L.J.; González, A.S.; Vega, V.; De la Prida, V.M. Electrostatic supercapacitors by atomic layer deposition on nanoporous anodic alumina templates for environmentally sustainable energy storage. *Coatings* **2018**, *8*, 403. [CrossRef]
19. Pavesi, L. Porous silicon dielectric multilayers and microcavities. *La Riv. Del Nuovo Cimento* **1997**, *20*, 1–76. [CrossRef]
20. Starkey, T.; Vukusic, P. Light manipulation principles in biological photonic systems. *Nanophotonics* **2013**, *2*, 289–307. [CrossRef]
21. Sulka, G.D.; Hnida, K. Distributed Bragg reflector based on porous anodic alumina fabricated by pulse anodization. *Nanotechnology* **2012**, *23*, 075303. [CrossRef] [PubMed]
22. Zhang, Y.; Fu, Q.; Ge, J. Photonic sensing of organic solvents through geometric study of dynamic reflection spectrum. *Nat. Commun.* **2015**, *6*, 7510. [CrossRef] [PubMed]
23. Yang, D.; Tian, H.; Ji, Y. Nanoscale photonic crystal sensor arrays on monolithic substrates using side-coupled resonant cavity arrays. *Opt. Express* **2011**, *19*, 20023–20034. [CrossRef]
24. Zhang, Y.; Zhao, Y.; Lv, R. A review for optical sensors based on photonic crystal cavities. *Sens. Actuators A* **2015**, *233*, 374–389. [CrossRef]
25. Martín-Palma, R.J.; Torres-Costa, V.; Pantano, C.G. Distributed Bragg reflectors based on chalcogenide glasses for chemical optical sensing. *J. Phys. D Appl. Phys.* **2009**, *42*, 055109. [CrossRef]
26. Lee, W.; Kim, J.C. Highly Ordered porous alumina with tailor-made pore structures by pulse anodization. *Nanotechnology* **2010**, *21*, 485304. [CrossRef]
27. Chung, C.K.; Zhou, R.X.; Liu, T.Y.; Chang, W.T. Hybrid pulse anodization for the fabrication of porous anodic alumina films from commercial purity (99%) aluminum at room temperature. *Nanotechnology* **2009**, *20*, 055301. [CrossRef]

28. Chen, Y.; Santos, A.; Wang, Y.; Kumeria, T.; Li, J.; Wang, C.; Losic, D. Biomimetic nanoporius anodic alumina distributed Bragg reflectors in the form of films and microsized particles for sensing applications. *ACS Appl. Mater. Interfaces* **2015**, *7*, 19816–19824. [CrossRef]
29. Santos, A.; Yoo, J.H.; Rohatgi, C.V.; Kumeria, T.; Wang, Y.; Losic, D. Realisation and advanced engineering of true optical rugate filters based on nanoporous anodic alumina by sinusoidal pulse anodization. *Nanoscale* **2016**, *8*, 1360–1373. [CrossRef]
30. Santos, A.; Law, C.S.; Pereira, T.; Losic, D. Nanoporous hard data: Optical encoding of information within nanoporous anodic alumina photonic crystals. *Nanoscale* **2016**, *8*, 8091–8100. [CrossRef]
31. Law, C.S.; Santos, A.; Nemati, M.; Losic, D. Structural engineering of nanoporous anodic alumina photonic crystals by sawtooth-like pulse anodization. *ACS Appl. Mater. Interfaces* **2016**, *8*, 13542–13554. [CrossRef] [PubMed]
32. Santos, A.; Pereira, T.; Law, C.S.; Losic, D. Rational engineering of nanoporous anodic alumina optical bandpass filters. *Nanoscale* **2016**, *8*, 14846–14857. [CrossRef] [PubMed]
33. Wang, B.; Fei, G.T.; Wang, M.; Kong, M.G.; Zhang, L.D. Preparation of photonic crystals made of air pores in anodic alumina. *Nanotechnology* **2007**, *18*, 365601. [CrossRef]
34. Wang, Y.; Chen, Y.; Kumeria, T.; Ding, F.; Evdokiou, A.; Losic, D.; Santos, A. Facile synthesis of optical microcavities by a rationally designed anodization approach: Tailoring photonic signals by nanopore structure. *ACS Appl. Mater. Interfaces* **2015**, *7*, 9879–9888. [CrossRef] [PubMed]
35. Kumeria, T.; Rahman, M.M.; Santos, A.; Ferré-Borrull, J.; Marsal, L.F.; Losic, D. Nanoporous Anodic alumina rugate filters for sensing of ionic mercury: Toward environmental point-of-analysis systems. *ACS Appl. Mater. Interfaces* **2014**, *6*, 12971–12978. [CrossRef] [PubMed]
36. Kumeria, T.; Rahman, M.M.; Santos, A.; Ferré-Borrull, J.; Marsal, L.F.; Losic, D. Structural and optical nanoengineering of nanoporous anodic alumina rugate filters for real-time and label-free biosensing applications. *Anal. Chem.* **2014**, *86*, 1837–1844. [CrossRef] [PubMed]
37. Lee, W.; Schwirn, K.; Steinhart, M.; Pippel, E.; Scholz, R.; Gösele, U. Structural engineering of nanoporous anodic aluminium oxide by pulse anodization of aluminium. *Nat. Nanotechnol.* **2008**, *3*, 234–239. [CrossRef]
38. Zhang, S.; Xu, Q.; Feng, S.; Sun, C.; Peng, Q.; Lan, T. The effect of the voltage waveform on the microstructure and optical properties of porous anodic alumina photonic crystals. *Opt. Mater.* **2019**, *98*, 109488. [CrossRef]
39. Li, S.-Y.; Wang, J.; Wang, G.; Wang, J.-Z.; Wang, C.-W. Fabrication of one-dimensional alumina photonic crystals by anodization using a modified pulse-voltage method. *Mater. Res. Bull.* **2015**, *68*, 42–48. [CrossRef]
40. An, Y.-Y.; Wang, J.; Zhou, W.-M.; Jin, H.-X.; Li, J.-F.; Wang, C.-W. The preparation of high quality alumina defective photonic crystals and their application of photoluminescence enhancement. *Superlattices Microstruct.* **2018**, *119*, 1–8. [CrossRef]
41. Abbasimofrad, S.; Kashi, M.A.; Noormohammadi, M.; Ramazani, A. Tuning the optical properties of nanoporus anodic alumina photonic crystals of allowed voltage range via mixed acid concentration. *J. Phys. Chem. Sol.* **2018**, *118*, 221–231. [CrossRef]
42. Zheng, W.J.; Fei, G.T.; Wang, B.; Zhang, L.D. Modulation of transmission spectra of anodized alumina membrane distributed Bragg reflector by controlling anodization temperature. *Nanoscale Res. Lett.* **2009**, *4*, 665–667. [CrossRef] [PubMed]
43. Ferré-Borrull, J.; Rahman, M.M.; Pallarès, J.; Marsal, L.F. Tuning nanoporous anodic alumina distributed-Bragg reflectors with the number of anodization cycles and the anodization temperature. *Nanoscale Res. Lett.* **2014**, *9*, 416. [CrossRef] [PubMed]
44. Lee, W.; Ji, R.; Gösele, U.; Nielsch, K. Fast fabrication of long-range ordered porous alumina membranes by hard anodization. *Nat. Mater.* **2006**, *5*, 741–747. [CrossRef] [PubMed]
45. Yi, L.; Zhiyuan, L.; Xing, H.; Yisen, L.; Yi, C. Formation and microstructures of unique nanoporous AAO films fabricated by high voltage anodization. *J. Mater. Chem.* **2011**, *21*, 9661–9666. [CrossRef]
46. Yi, L.; Zhiyuan, L.; Xing, H.; Yisen, L.; Yi, C. Investigation of intrinsic mechanisms of aluminium anodization processes by analysing the current density. *RSC Adv.* **2012**, *2*, 5164–5171. [CrossRef]
47. Ozin, G.A.; Arsenault, A. *Nanochemistry: A Chemical Approach to Nanomaterials*, 2nd ed.; Royal Society of Chemistry: Cambridge, UK, 2015.
48. Byrnes, J. *Unexploded Ordnance Detection and Mitigation*, 1st ed.; Springer: Heidelberg, Germany, 2009; pp. 21–22.

49. Włodarski, M.; Putkonen, M.; Norek, M. Infrared absorption study of Zn-S hybrid and ZnS ultrathin films deposited on poorus AAO ceramic support. *Coatings* **2020**, *10*, 459. [CrossRef]
50. Zhang, L.; Dong, K.; Chen, D.; Liu, Y.; Xue, J.; Lu, H.; Zhang, R.; Zheng, Y. Solar-blind ultraviolet AlInN/AlGaN distributed Bragg reflectors. *Appl. Phys. Lett.* **2013**, *102*, 242112. [CrossRef]

© 2020 by the authors. Licensee MDPI, Basel, Switzerland. This article is an open access article distributed under the terms and conditions of the Creative Commons Attribution (CC BY) license (http://creativecommons.org/licenses/by/4.0/).

Article

A Proposal for a Composite with Temperature-Independent Thermophysical Properties: HfV$_2$–HfV$_2$O$_7$

Philipp Keuter [1,*], Anna L. Ravensburg [1,2], Marcus Hans [1], Soheil Karimi Aghda [1], Damian M. Holzapfel [1], Daniel Primetzhofer [2] and Jochen M. Schneider [1]

[1] Materials Chemistry, RWTH Aachen University, Kopernikusstr. 10, 52074 Aachen, Germany; anna.ravensburg@physics.uu.se (A.L.R.); hans@mch.rwth-aachen.de (M.H.); karimi@mch.rwth-aachen.de (S.K.A.); holzapfel@mch.rwth-aachen.de (D.M.H.); schneider@mch.rwth-aachen.de (J.M.S.)
[2] Department of Physics and Astronomy, Uppsala University, Box 516, 75120 Uppsala, Sweden; daniel.primetzhofer@physics.uu.se
* Correspondence: keuter@mch.rwth-aachen.de

Received: 9 October 2020; Accepted: 5 November 2020; Published: 7 November 2020

Abstract: The HfV$_2$–HfV$_2$O$_7$ composite is proposed as a material with potentially temperature-independent thermophysical properties due to the combination of anomalously increasing thermoelastic constants of HfV$_2$ with the negative thermal expansion of HfV$_2$O$_7$. Based on literature data, the coexistence of both a near-zero temperature coefficient of elasticity and a coefficient of thermal expansion is suggested for a composite with a phase fraction of approximately 30 vol.% HfV$_2$ and 70 vol.% HfV$_2$O$_7$. To produce HfV$_2$–HfV$_2$O$_7$ composites, two synthesis pathways were investigated: (1) annealing of sputtered HfV$_2$ films in air to form HfV$_2$O$_7$ oxide on the surface and (2) sputtering of HfV$_2$O$_7$/HfV$_2$ bilayers. The high oxygen mobility in HfV$_2$ is suggested to inhibit the formation of crystalline HfV$_2$–HfV$_2$O$_7$ composites by annealing HfV$_2$ in air due to oxygen-incorporation-induced amorphization of HfV$_2$. Reducing the formation temperature of crystalline HfV$_2$O$_7$ from 550 °C, as obtained upon annealing, to 300 °C using reactive sputtering enables the synthesis of crystalline bilayered HfV$_2$–HfV$_2$O$_7$.

Keywords: thermoelasticity; negative thermal expansion; composites; magnetron sputtering

1. Introduction

Volume expansion upon heating is probably the most prominent example of the influence of temperature on the physical properties of materials. However, for electronic, optical, and other high-precision devices, whose performances are critically affected by slight variations in volume, near-zero thermal expansion materials are desired [1–5]. Moreover, for mechanical components of precision instruments, temperature-independent volumes and elastic moduli are required [6–8]. A combination of both properties has only been obtained in gum metals [9,10] and Fe-Ni alloys [11] after intense deformation, thereby promoting the development of materials with intrinsic near-zero expansion and temperature-invariant elastic behavior irrespective of the plastic deformation route.

To compensate for thermally-induced volume changes, implementation of materials with negative thermal expansion (NTE) in a composite is a widely propagated approach [4,5,12]. Here, we propose to join a material with NTE properties with another exhibiting anomalously increasing thermoelastic constants in an attempt to obtain a combination of temperature-invariant elastic behavior and volume. This is a pioneering approach for the design of materials with temperature-independent thermophysical properties.

However, elastic constants commonly decrease monotonically with rising temperatures due to the anharmonicity of lattice vibrations [13,14]. Materials that deviate from this trend, thereby exhibiting anomalous thermoelastic behavior, constitute promising candidates for the design of composites with temperature-independent thermoelastic properties. While nowadays NTE has been observed in a wide variety of material families, e.g., zirconium tungstates and vanadates, zeolites, metal cyanides, metal-organic framework materials, perovskites, and anti-perovskites [12,15–18], reports of materials with anomalous thermoelastic behavior are scarce. The cubic transition metals V, Nb, Ta, Pd, and Pt each exhibit a thermoelastic anomaly in their shear elastic constants c_{44} [19–23]. Furthermore, binary Nb–Zr, Nb–Mo, Pd–Ag, and Pd–Rh solid solutions behave anomalously within well-defined concentration ranges [24–26]. Experimental and theoretical studies revealed the combination of a high density of states and electronic reallocation upon lattice distortion in the vicinity of the Fermi level to be the physical origin of the anomalous thermoelastic behavior [27–31]. However, the anomaly in these systems is mostly limited to the shear elastic constant c_{44}, whereas the remaining elastic constants (there are three independent elastic constants for cubic symmetries, i.e., c_{11}, c_{12}, and c_{44}) behave normally. Consequently, their thermoelastic anomaly is highly directionally dependent, but an isotropic temperature-independent behavior is desired for the proposed composite. Intermetallic cubic HfV_2 (space group: Fm-3m), which is stable from around 112 K up to the melting point of 1820 K [32,33], exhibits increasing thermoelastic constants in all probed directions [32] so that an increase in the macroscopic elastic modulus (E) upon heating was measured in polycrystalline samples [34,35]. Thus, HfV_2 is an ideal constituent to aim for a temperature-invariant elastic behavior in a composite. The second component, consequently, serves then for the compensation of the thermoelastic increase and for the positive thermal expansion (PTE) of HfV_2. The linear coefficient of thermal expansion of HfV_2 has been measured to be 9.9×10^{-6} K^{-1} around room temperature [36].

Multiple mechanisms may give rise to NTE [5,17], e.g., the magnetovolume effect, phase transitions, atomic radius contraction, and flexible network structures, whereas the latter is the prevalent physical origin in most NTE materials. In general, the expansion effect due to longitudinal vibrations in these flexible network structures is over-compensated by the contraction owing to transverse vibrations [5,37]. In regard to the search for suitable NTE materials, anisotropic contraction restricts their practical usability, so cubic phases, which exhibit inherently isotropic contraction, are particularly promising. In the group of flexible network materials, ZrW_2O_8 and isostructural HfW_2O_8 have an unprecedented isotropic NTE range of 0.3 to 1050 K [38,39]. Besides, the only ternary oxide within the Hf-V-O system, HfV_2O_7, exhibits isotropic NTE above approximately 370 K with a negative coefficient of thermal expansion of -7.2×10^{-6} K^{-1} [40]. With decreasing temperature, phase transformations first into an incommensurate structure, stable between around 369 and 340 K, and finally into a cubic $3 \times 3 \times 3$ superstructure, are obtained [41]. Hence, forming a composite of HfV_2 with the corresponding ternary oxide, HfV_2O_7, appears promising for tailoring the physical properties of the composite material towards thermal invariance.

Physical vapor deposition techniques have proven to be successful in synthesizing and refining materials with NTE properties [42–45]. Consequently, after assessing that a combination of both temperature-invariant elastic behavior and volume is achieved simultaneously for a certain phase fraction ratio, two synthesis pathways for HfV_2–HfV_2O_7 composites were studied: (1) annealing of magnetron sputtered HfV_2 thin films in air to form a HfV_2O_7 oxide scale on the thin film surface and (2) magnetron sputtering of HfV_2O_7 on HfV_2. Furthermore, the NTE properties of HfV_2O_7 were verified for single-layered HfV_2O_7 thin films using temperature-dependent in situ X-ray diffraction.

2. Materials and Methods

2.1. Experimental Methods

Depositions were carried out by direct current magnetron sputtering at a target-to-substrate distance of 10 cm using elemental Hf and V (both 99.9% purity, 50 mm diameter) targets while the substrate remained at floating potential. For the synthesis of stoichiometric HfV_2, the employed target power densities were 2.0 and 10.2 W cm^{-2} for Hf and V, respectively. Ar (99.9999%) was used as sputtering gas to achieve a working pressure of 0.4 Pa. The base pressure (at deposition temperature) was below 1×10^{-6} Pa. HfV_2 films were deposited without intentional heating and at substrate temperatures of 500 and 700 °C. To ensure high purity of deposited HfV_2, the following measures were taken: The backsides of the single-crystalline sapphire (0001) substrates were deposited beforehand with approximately 250 nm of Nb to overcome their partial transmissivity to radiation during heating. This consequently reduced the heat impact on the chamber walls, and thus, a considerable decrease in base pressure at elevated deposition temperatures was achieved. Second, prior to each deposition, all targets were sputtered for 5 min with closed shutters (positioned approximately 2 cm opposite of the target) to make the surfaces free from condensed impurities and to getter residual gases. Third, elemental Zr was additionally sputtered at 20 W against a shutter during all depositions to exploit its pronounced affinity for oxygen [46,47] as a getter pump to further reduce residual gas incorporation [48] into the growing HfV_2 thin film. The Zr concentration in the as-deposited films was below 0.7 at.% based on energy-dispersive X-ray spectroscopy (EDX). Subsequently, selected films were capped after cooling to room temperature with an approximately 10 nm thick Nb layer to prevent impurity incorporation into the as-deposited thin film during air exposure.

HfV_2O_7 thin films were deposited at a substrate temperature of 450 °C in a reactive Ar/O_2 (99.999%) atmosphere at a constant working pressure of 0.86 Pa while varying the O_2 partial pressure between 0.05 and 0.13 Pa. Further information can be found elsewhere [49]. For the HfV_2–HfV_2O_7 bilayer deposition, the synthesis procedure was the following: First, HfV_2 was deposited at 700 °C for 75 min. Afterwards, the system was cooled down in vacuum to the synthesis substrate temperatures for HfV_2O_7, which were 250, 300, and 350 °C. Pure Ar was used for plasma ignition before O_2 (p_{O2} = 0.09 Pa) was introduced to reactively sputter HfV_2O_7 for 100 min.

Annealing experiments were performed in a GERO F 40-200/13 air furnace (Carbolite Gero, Neuhausen, Germany). The Hf-V ratio in the synthesized films was measured by EDX carried out in a JEOL JFM-6480 SEM (JEOL Ltd., Tokyo, Japan) equipped with an EDAX Genesis 2000 device (EDAX Inc., Mahwah, NJ, USA) at an acceleration voltage of 20 kV. Chemical composition depth profiling of HfV_2 was done by time-of-flight elastic recoil detection analysis (ToF-ERDA) at the Tandem Accelerator Laboratory of Uppsala University. 36 MeV $^{127}|^{8+}$ primary ions and a time-of-flight telescope in combination with a Si solid-state detector for energy discrimination were used. Further details on the detector telescope can be found elsewhere [50]. The time and energy coincidence spectra were evaluated using the CONTES software package [51]. O and H depth profiles were characterized by ToF-ERDA, while the Hf-V ratio was obtained from EDX spectra. It should be noticed that Nb-capped und uncapped HfV_2 were measured by ToF-ERDA under identical conditions, hence, systematic uncertainties do not affect this comparison. The structure of the synthesized films was studied using X-ray diffraction (XRD) carried out with a Bruker AXS D8 Discover General Area Detector Diffraction System (Bruker Corporation, Billerica, MA, USA). A Cu K$_\alpha$ source (current 40 mA, voltage of 40 kV) was used with a 0.5 mm pinhole collimator. Scans were performed at a fixed incidence angle of 15°. Selected samples were measured between room temperature and 475 °C using a DHS 1100 Hot Stage (Anton Paar, Ostfildern-Scharnhausen, Germany) equipped with a NI-NiCr thermocouple to measure the surface temperature. Peak fitting was conducted using TOPAS software (version 3) with a pseudo-Voigt II function. Lattice parameters of the cubic structures were consequently calculated employing Bragg's law [52]. The lattice parameters of HfV_2 were determined by averaging the data from (220), (311), (222), (331), (422), (333), and (044) reflections. Based on the obtained changes in lattice parameters with temperature, the linear coefficient of thermal expansion for cubic HfV_2O_7 was calculated by averaging the data from (200), (210), (211), (211), (220), (311), and (222) reflections. Standard deviations are added to evaluate the fitting quality. Morphology of bilayered HfV_2–HfV_2O_7 was studied using scanning transmission electron microscopy (STEM) carried out in an FEI Helios Nanolab 660 dual-beam microscope (Thermo Fisher Scientific, Waltham, MA, USA). Cross-sectional sample preparation was conducted by focused ion beam techniques with a Ga$^+$ source following a standard lift-out procedure [53].

2.2. Theoretical Methods

Ground state equilibrium lattice parameters for cubic HfO$_2$ (Fm-3m, 12 atoms), VO$_2$ (Fm-3m, 12 atoms), VO (Fm-3m, 8 atoms), and HfO (Fm-3m, 8 atoms) were calculated within the framework of density functional theory (DFT) [54] employing the Vienna ab initio simulation package (VASP) [55,56] with projector augmented wave potentials. The generalized gradient approximation, as introduced by Perdew, Burke, and Ernzerhof [57], was used for all calculations. Integration in the Brillouin zone was performed on a 20 × 20 × 20 k-point grid according to Monkhorst and Pack [58]. The total energy convergence criterion was 0.01 meV within a 500 eV cut-off. The equilibrium lattice parameters were determined by a Birch–Murnaghan equation of state [59,60] fit.

3. Results and Discussion

3.1. Composite Assessment

First, the suitability of HfV$_2$–HfV$_2$O$_7$ as a composite with temperature-independent physical properties is evaluated, which relies on the coexistence of a near-zero coefficient of thermal expansion and a near-zero temperature coefficient of elasticity (TCE) for a certain phase fraction ratio. The TCE is defined by

$$TCE = \frac{1}{E}\frac{dE}{dT} \quad (1)$$

and is estimated for HfV$_2$–HfV$_2$O$_7$ composite using a rule-of-mixture approach (weighted average based on the volume fractions). While the temperature dependence of the elastic modulus of HfV$_2$ is taken from the literature [35], no thermoelasticity data for HfV$_2$O$_7$ have been reported. The temperature-dependent elastic modulus has consequently been estimated from the experimentally obtained elastic modulus of HfV$_2$O$_7$ [49] assuming the same relative decline with temperature, as reported for the NTE material ZrW$_2$O$_8$ [61]. ZrW$_2$O$_8$ and HfV$_2$O$_7$ exhibit comparable negative coefficients of thermal expansion of -9.1×10^{-6} [62] and -7.2×10^{-6} K^{-1} [40], respectively, and share common structural features, both forming an openly-packed network structure of octahedral (Zr/Hf)O$_6$ and polyhedral (W/V)O$_4$ units connected by corner-sharing oxygen atoms [63]. The TCE of the composite was averaged over a temperature range from 120 (onset of NTE behavior in HfV$_2$O$_7$ [40,41,64]) to 300 °C.

On the other hand, the coefficient of thermal expansion of a composite usually does not follow a simple rule-of-mixture behavior [65]. NTE materials are generally less stiff (lower E) than expected based on the bond strengths [66], and thus normally constitute the more compliant component in the composite. As a result, the elastic deformation during expansion and contraction is predominantly concentrated on the NTE component reducing its impact on the overall expansion coefficient. Consequently, theoretical models to describe the expansion coefficient of a composite typically also take the elastic properties of the individual components into account [66]. In the Turner model the overall coefficient of thermal expansion of the composite α_c is described by

$$\alpha_c = \frac{\sum_i B_i \alpha_i \phi_i}{\sum_i B_i \phi_i}, \quad (2)$$

where B_i, α_i, and ϕ_i denote the bulk modulus, coefficient of thermal expansion, and volume fraction, respectively, of component i [65]. The bulk modulus and the coefficient of thermal expansion of HfV$_2$ (HfV$_2$O$_7$) of 117 GPa [32] (56 GPa [49]) and 9.9×10^{-6} K^{-1} [36] (-7.2×10^{-6} K^{-1} [40]), respectively, were used for the calculation of α_c. The resulting volume fraction dependent expansion coefficient α_c and averaged TCE of HfV$_2$–HfV$_2$O$_7$ composite are plotted in Figure 1.

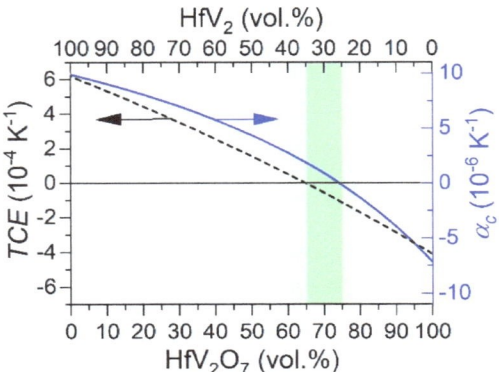

Figure 1. Variation of the temperature coefficient of elasticity TCE (black dashed line) and linear coefficient of thermal expansion α_c (blue solid line) of composite material HfV_2–HfV_2O_7 depending on the volume fractions of the individual constituents. The region of interest is highlighted in green.

The coefficient of thermal expansion of the composite was calculated to reach temperature invariance ($\alpha_c = 0$) at a volume fraction of HfV_2 of around 0.25 ($\phi_{HfV2O7} = 0.75$). A TCE of zero was achieved for ϕ_{HfV2} of around 0.35 ($\phi_{HfV2O7} = 0.65$) with a corresponding α_c of 1.9×10^{-6} K^{-1}, which complies with the class of very low thermal expansion materials [1]. These results not only suggest that α_c and TCE of this composite can individually be adjusted to thermal invariance but also that a near-zero α_c and TCE may be achieved simultaneously. Based on the applied data and the assumptions outlined above, we predict that the corresponding volume phase fractions of HfV_2 (HfV_2O_7) are between 20 and 40 (80–60) vol.% depending on the optimization criterion. After assessing the capability of composite HfV_2–HfV_2O_7 to exhibit temperature-independent properties, its synthesis is studied in the following.

3.2. Composite Formation by Oxidation of HfV_2

A potential synthesis pathway for a composite has been demonstrated by oxidizing TiN thin films, where the time- and temperature-dependent formation of a TiO_2 scale on top of TiN was observed with varying thickness, and hence phase fraction ratios [67,68]. Consequently, the first synthesis strategy to form HfV_2–HfV_2O_7 composites comprises synthesis of HfV_2 and subsequent heat treatment in air to partly oxidize HfV_2 in an attempt to form a HfV_2O_7 oxide scale on top.

3.2.1. Phase Formation of HfV_2

First, the phase formation for co-sputtered Hf-V thin films using magnetron sputtering is described. No previous reports on the synthesis of HfV_2 thin films by magnetron sputtering are currently available. However, the synthesis of isoelectronic and isostructural ZrV_2 thin films is discussed in the literature [69–71]. EDX analysis of the as-deposited thin films shows deviations from the desired 1Hf:2V stoichiometry below ±1 at.%. The results of the structural analysis to study the influence of the deposition temperature on the phase formation are depicted in Figure 2.

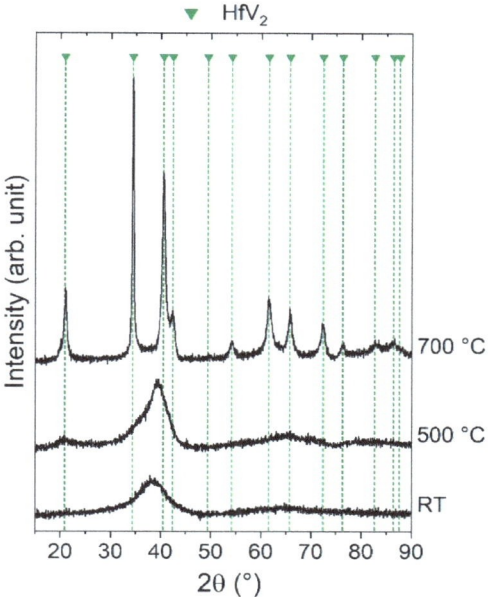

Figure 2. X-ray diffractograms of stoichiometric (V/Hf = 2.0) Hf-V samples synthesized without intentional heating (RT) and at synthesis temperatures of 500 and 700 °C.

A broad hump around 40° was obtained for HfV$_2$ deposited without intentional heating (RT), indicating an amorphous structure, as reported for magnetron sputtered Zr–V thin films deposited without heating [71] and at 400 °C [70]. In this temperature regime, in which the adatom mobility is low, the formation of amorphous solid solution in Zr–V can be understood based on its higher stability compared to a random bcc solid solution, as demonstrated by DFT calculations [71]. It is reasonable to assume that this may also apply to Hf-V.

For Zr–V, a phase formation sequence from amorphous (400 °C) to a phase mixture of bcc V and hcp Zr (500 °C) to intermetallic ZrV$_2$ (600 °C) was obtained with increasing deposition temperatures [70]. The same sequence was obtained by annealing amorphous films [70]. The intermediate formation of bcc V and hcp Zr is contradictive to the accepted equilibrium phase diagram and is mostly discussed with respect to difficult nucleation kinetics of the Laves phase structure [69,70]. However, it may also be explained by the thermodynamic instability of cubic ZrV$_2$, as predicted by DFT in the ground state [71–73], exhibiting an energy of formation of 150 meV atom^{-1} [71], which persists at elevated temperatures. In comparison, several theoretical studies also predict cubic HfV$_2$ to be energetically unstable in the ground state, reporting energies of formation between 20 and 35 meV per atom [29,74,75]. Furthermore, experiments show a transformation of cubic HfV$_2$ upon cooling into an orthorhombic structure at around −160 °C [76,77] possibly due to kinetically limited decomposition into elemental V and Hf, since orthorhombic HfV$_2$ also exhibits positive energy of formation [29,75]. However, for sputtered HfV$_2$ at 500 °C (see Figure 2), the change in shape of the main peak at around 38° suggests the first formation of nanocrystals, and the emerging hump around 20° points towards intermetallic HfV$_2$ nanocrystals, since their presence cannot be explained by hcp Hf or bcc V.

A further increase in deposition temperature to 700 °C resulted in the formation of sharp diffraction peaks which all can be assigned to cubic HfV$_2$ [78]. Thus, no phase mixture of hcp Hf and bcc V, unlike for Zr–V, was observed, but a sequence from amorphous to crystalline HfV$_2$ with increasing deposition temperature was obtained. Consequently, the high temperature of 700 °C required to form crystalline HfV$_2$ is attributed to the kinetically limited formation of the Laves phase structure and not to an energetic instability of the cubic structure up to these temperatures. This notion is supported by theoretical predictions suggesting the energetic stabilization of cubic HfV_2 at temperatures as low as −120 °C due to lattice vibrations [29]. For ZrV_2, due to the considerably higher energy of formation in the ground state [71], its energetic stabilization is expected at higher temperatures, potentially explaining the discussed phase formation differences between Hf-V and Zr–V thin films. For comparison, bulk synthesis of HfV_2 includes heat treatments at temperatures above 1200 °C [34,77,79]. The reduction in synthesis temperature to 700 °C is enabled by surface diffusion of adatoms during sputtering [80].

No indications for impurity phases based on the presented XRD results were obtained in these samples, which is ascribed to the measures outlined above. However, thin films synthesized at higher base pressures or without additional co-sputtering of Zr contained traces of Hf_3V_3O and HfO_2 (not shown). This is in agreement with the observation of ZrV_3O_3 as an impurity phase in sputtered Zr–V thin films [70].

3.2.2. Stability of HfV_2

As a next step, the stability of the synthesized HfV_2 films upon air exposure is examined. For this purpose, the structure of uncapped HfV_2 was studied as a function of the cumulated air-exposure time after removing it from the high-vacuum deposition system. The results are summarized in Figure 3.

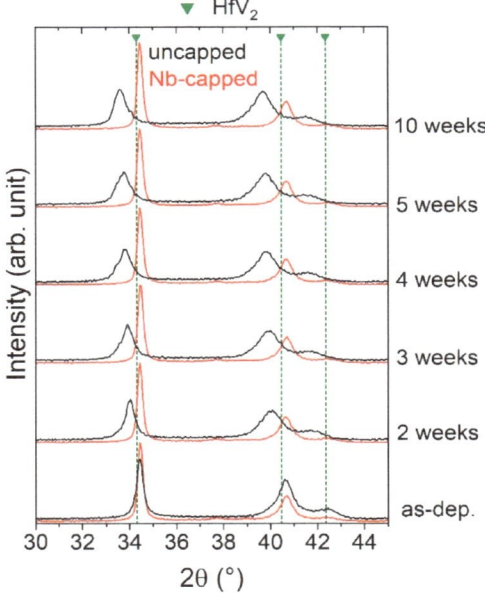

Figure 3. X-ray diffractograms of uncapped HfV_2 (black) and Nb-capped HfV_2 (red) for varying storage times in air.

In the as-deposited state, referring to a minimized air-exposure time of approximately 15 min, the peak positions coincide well with literature data for cubic HfV_2 [78]. With increasing air-exposure time, a continuous peak shift to lower 2θ values is obtained, indicating an increase in the lattice parameter of the cubic structure from 7.38 to 7.51 Å (+1.8%) after four and to 7.54 Å (+2.2%) after ten weeks in air (standard deviations ≤ 0.01 Å). The obtained increase in lattice parameter may suggest continuous interstitial incorporation of impurities into the HfV_2 lattice, which consequently also explains reported property changes in $(Hf,Zr)V_2$ bulk samples after a one-year storage period [81]. HfV_2 has been reported to act as a strong (weak) getter for hydrogen (oxygen) by dissociation of water [82].

Thin metal capping layers were shown to serve as effective oxidation barriers during air exposure at room temperature [83]. Hence, as-deposited HfV$_2$ was capped with approximately 10 nm Nb due to its passivating properties [84]. The functionality of this capping layer was investigated systematically by comparing the stabilities of capped and uncapped HfV$_2$ using XRD (see Figure 3). Other than a small additional peak around 38°, measured for the Nb-capped film, which is attributed to bcc Nb, no difference in phase composition between both HfV$_2$ films in the as-deposited state was observed. However, in contrast to the uncapped HfV$_2$ film, no peak shift with increasing air-exposure time was obtained, demonstrating that the Nb capping ensures protection of synthesized HfV$_2$ against the incorporation of impurities from the ambient at room temperature.

To identify the incorporated impurities, the chemical compositions of both samples, uncapped and Nb-capped, were analyzed four weeks after deposition using ToF-ERDA. The measured oxygen concentration as a function of the film thicknesses is presented in Figure 4. The depth scale was calculated with the atomic masses of Hf and V under the assumption of stoichiometric HfV$_2$ with a density of 9.3 g cm^{-3} [32].

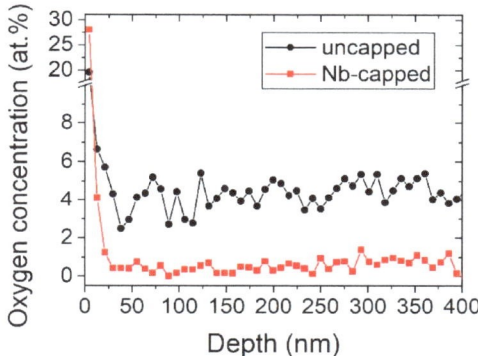

Figure 4. ToF-ERDA depth profile of the oxygen concentration for uncapped (black circle) and Nb-capped (red square) HfV$_2$ after four weeks of air-exposure.

Uncapped HfV$_2$ exhibited an averaged oxygen concentration of 4.3 ± 0.7 at.% in the bulk part of the film (neglecting the top surface oxidation) and a minor amount of hydrogen (<0.3 at.%, not shown) while the average oxygen concentration in the Nb-capped film was 0.5 ± 0.3 at.% and hydrogen was below the detection limit. Thus, the increasing lattice parameter for uncapped HfV$_2$ is primarily attributed to a continuous interstitial uptake of oxygen into the cubic structure upon exposure to air. It has been shown theoretically that interstitially incorporated oxygen contributes to the energetic and mechanical stabilization of the cubic structure [29]. The oxygen concentration was measured to be constant throughout the analyzed film thickness of approximately 400 nm, suggesting high mobility of oxygen in cubic HfV$_2$ already at room temperature. This may also have implications for the oxidation behavior of these films at elevated temperatures, which is discussed in the following.

3.2.3. Oxidation of HfV$_2$

One sample of uncapped HfV$_2$ was cyclically annealed in air for approximately 30 min at temperatures between 150 and 650 °C in 100 °C intervals. After each annealing step, the structure of the sample was studied by XRD. The resulting diffractograms are shown in Figure 5.

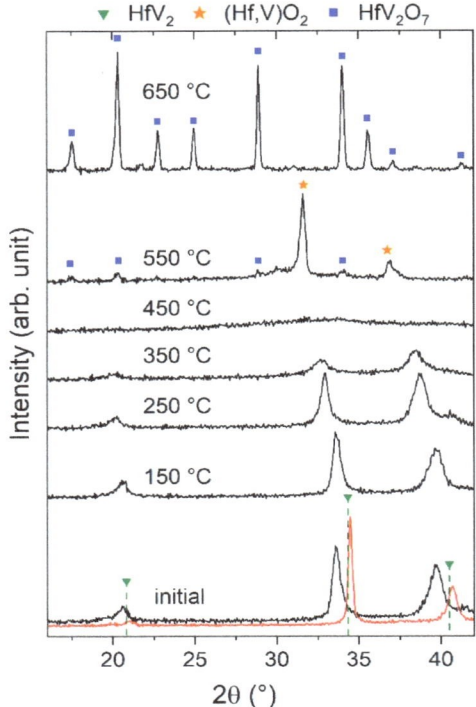

Figure 5. X-ray diffractograms of an uncapped HfV$_2$ thin film (9 weeks after deposition) annealed at the indicated temperatures for approximately 30 min. For comparison, the diffractogram of a Nb-capped film was added (red).

In the initial state, the sample was already exposed to air for 9 weeks. As a consequence, peak positions were already shifted towards smaller 2θ values compared to the as-deposited (see Figure 3) and Nb-capped films (see Figure 5). For all annealing steps up to 350 °C, the structural development is characterized by a continuous peak shift and broadening suggesting the successive incorporation of oxygen into the cubic structure finally yielding amorphization. as observed at 450 °C. Complementary annealing experiments with additional samples at fixed temperatures of 300 and 400 °C indicate that this process is determined by kinetics and may occur already at lower temperatures (not shown).

After annealing at 550 °C, small peaks emerged which may already indicate the formation of HfV$_2$O$_7$ nanocrystals [41], while the two main peaks at 32° and 37° cannot be attributed to any reported equilibrium phase in the Hf-V-O materials system. The peak positions of the unknown phase fit reasonably well to a cubic (fcc-based) structure exhibiting a lattice parameter of 4.9 Å. The reported cubic structures in the Hf-V-O systems, VO (Fm$\overline{3}$m) [85] and HfO$_2$ (Fm$\overline{3}$m) [86], can be excluded due to the difference in the lattice parameters. Furthermore, the formation of a hypothetic cubic Hf$_{1-x}$V$_x$O monoxide appears unlikely based on conducted DFT studies, predicting a lattice parameter between 4.19 and 4.58 Å for $x = 0$ and $x = 1$, respectively. Nevertheless, the experimentally obtained lattice parameter lies within the predicted lattice parameter range of a previously unreported Hf$_{1-x}$V$_x$O$_2$ dioxide (Fm$\overline{3}$m) which is between 5.08 and 4.74 Å for $x = 0$ and $x = 1$, respectively. Hence, the formation of a metastable ternary (Hf,V)O$_2$ phase is assumed henceforth. A high V solubility in cubic HfO$_2$ has also been reported elsewhere [87]. At 650 °C, the formation of XRD phase pure HfV$_2$O$_7$ is evident [41].

The observation that the first crystallites of HfV$_2$O$_7$ may form at temperatures of 550 °C while oxygen is highly mobile in HfV$_2$ already at room temperature suggests that the formation of HfV$_2$–HfV$_2$O$_7$ composites is not feasible using conventional annealing experiments in air. The oxygen mobility in HfV$_2$ at the formation temperature of crystalline HfV$_2$O$_7$ appears to be critical for the endeavor of HfV$_2$–HfV$_2$O$_7$ composite formation. However, lowering of the synthesis temperature of crystalline HfV$_2$O$_7$ by using non-equilibrium-based synthesis approaches appears promising and is discussed in the following.

3.3. Composite Formation by Sputtering of Bilayered HfV$_2$–HfV$_2$O$_7$

3.3.1. Phase Formation of HfV$_2$O$_7$

The formation of crystalline HfV$_2$O$_7$ at a substrate temperature of 350 °C using reactive magnetron sputtering has been demonstrated previously [49], but the effect of the O$_2$ partial pressure has not been investigated yet. However, for the synthesis of bilayered HfV$_2$–HfV$_2$O$_7$ by magnetron sputtering, the O$_2$ partial pressure during the synthesis of HfV$_2$O$_7$ is expected to be decisive for the purity of the underlying HfV$_2$ film, due to its high oxygen affinity. Hence, the influence of the O$_2$ partial pressure on the phase formation of HfV$_2$O$_7$ was studied systematically for a synthesis temperature of 450 °C. The results of the structural analysis of synthesized thin films, exhibiting a V–Hf ratio of 2.0 based on EDX, are shown in Figure 6.

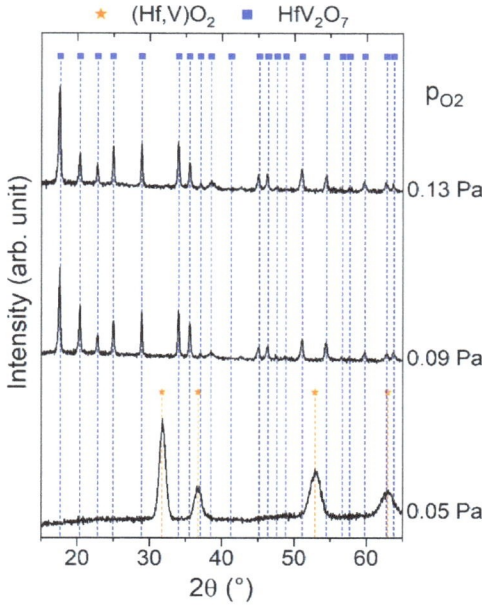

Figure 6. X-ray diffractograms of reactively sputtered Hf-V–O (V/Hf = 2) thin films deposited at 450 °C with varying O$_2$ partial pressure.

The thin film sputtered at an O$_2$ partial pressure of 0.05 Pa exhibits broad peaks at around 32°, 37°, 53°, and 63°. The peak positions coincide well with the ones of the cubic (Hf,V)O$_2$ structure observed during annealing of HfV$_2$ at 550 °C (see Figure 5). Cubic HfO$_2$ usually constitutes the high-temperature polymorph, which is stable above 2600 °C at ambient pressure [86] but can be stabilized down to room temperature by V doping in specific atmospheres (oxidation state ≤ V^{4+}) [88]. Furthermore, cubic HfO$_2$ thin films form by ion beam assisted deposition under oxygen deficiency and substrate cooling [89]. Thus, it is assumed that the formation of cubic (Hf,V)O$_2$ obtained for the lowest O$_2$ partial pressure is triggered by oxygen deficiency. This notion is supported by the fact that the increase in the O$_2$ partial pressure from 0.05 to 0.09 Pa resulted in a sixfold decrease in deposition rate (with respect to the mass gain), which indicates a more pronounced poisoning of the target racetrack. For O$_2$ partial pressures of 0.09 and 0.13 Pa, all obtained peaks can be assigned to the HfV_2O_7 structure [41], indicating the formation of phase pure HfV_2O_7 based on XRD.

3.3.2. Thermal Expansion of Sputtered HfV_2O_7

While numerous NTE materials have intensively been studied in bulk, studies on thin films are scarce [42–45]. To verify the NTE behavior of the synthesized HfV_2O_7 thin films, temperature-dependent in situ XRD measurements were performed to measure the change in the lattice parameter upon heating. The results are shown in Figure 7.

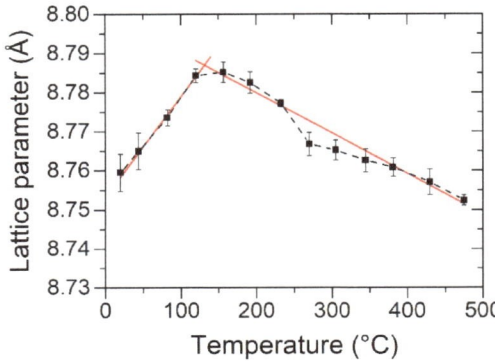

Figure 7. Lattice parameter of synthesized cubic HfV_2O_7 as a function of annealing temperature obtained by X-ray diffraction. A linear fit within the positive and negative thermal expansion range, respectively, was added (red).

A continuous increase in the lattice parameter between 20 and 120 °C was measured, whereas between 155 and 475 °C the lattice parameter decreased, indicating NTE. The obtained transition from PTE to NTE is in agreement with reported transition temperatures of HfV_2O_7 being around 120–130 °C produced by solid-state reaction synthesis approaches [40,41,64]. Additional wafer curvature measurements confirm this transition, as indicated by a change from increasing compressive to increasing tensile stress upon heating (not shown). The NTE properties of HfV_2O_7 originate from its openly-packed network structure consisting of octahedral HfO_6 and tetrahedral VO_4 units, constituting quasi-rigid building blocks that are interconnected by corner-sharing oxygen atoms [4,90,91]. Three of four oxygen atoms of the VO_4 tetrahedron are shared with neighboring HfO_6 octahedra, while one is shared with another VO_4 tetrahedron, thereby forming a V_2O_7 group [92]. The lattice symmetry (space group: $Pa\bar{3}$) restricts these O_3V–O–VO_3 bonds to retain an angle of 180°, allowing for transverse vibrations that may give rise to NTE [4,93]. The loss of the NTE properties at lower temperatures is related to structural transitions. HfV_2O_7 transforms upon cooling via an intermediate incommensurate structure to a $3 \times 3 \times 3$ superstructure [41,64]. It has been shown for isostructural ZrV_2O_7 that the majority of the O_3V–O–VO_3 linkages bend away from 180° in the $3 \times 3 \times 3$ superstructure [94–96]. However, the additional weak reflections of the HfV_2O_7 superstructure [97] could not be resolved by XRD in this work. Based on the presented data, the linear coefficients of thermal expansion were determined to be $(2.8 \pm 0.5) \times 10^{-5}$ K^{-1} (20 °C $\leq$ T $\leq$ 120 °C) and $(-9.9 \pm 0.9) \times 10^{-6}$ K^{-1} (155 °C $\leq$ T $\leq$ 475 °C), respectively, agreeing reasonably well with the reported linear coefficients of thermal expansion of 2.5×10^{-5} K^{-1} [64] and -7.2×10^{-6} K^{-1} [40]. After establishing a synthesis recipe for NTE material HfV_2O_7 that is characterized by a low required O_2 partial pressure, the way is paved for the synthesis of bilayered HfV_2–HfV_2O_7 composites using a two-staged sputtering process.

3.3.3. Phase Formation of HfV$_2$–HfV$_2$O$_7$ Bilayers

Besides a low O$_2$ partial pressure during reactive sputtering of HfV$_2$O$_7$, the synthesis temperature for HfV$_2$O$_7$ is decisive for avoiding substantial oxygen incorporation into HfV$_2$. Hence, the influence of synthesis temperature of HfV$_2$O$_7$, which was varied between 250 and 350 °C, on the phase formation of HfV$_2$–HfV$_2$O$_7$ composite was investigated by maintaining a low O$_2$ partial pressure of 0.09 Pa (see Figure 6). While a Nb passivation layer ensures protection of HfV$_2$ in air at room temperature (see Figure 3), additional bilayer depositions revealed that the deposition of a Nb interlayer, separating the two layers in the composite, does not prevent oxygen incorporation in HfV$_2$ during reactive sputtering of HfV$_2$O$_7$ (not shown) and is therefore not discussed further. The results of the structural analysis are shown in Figure 8.

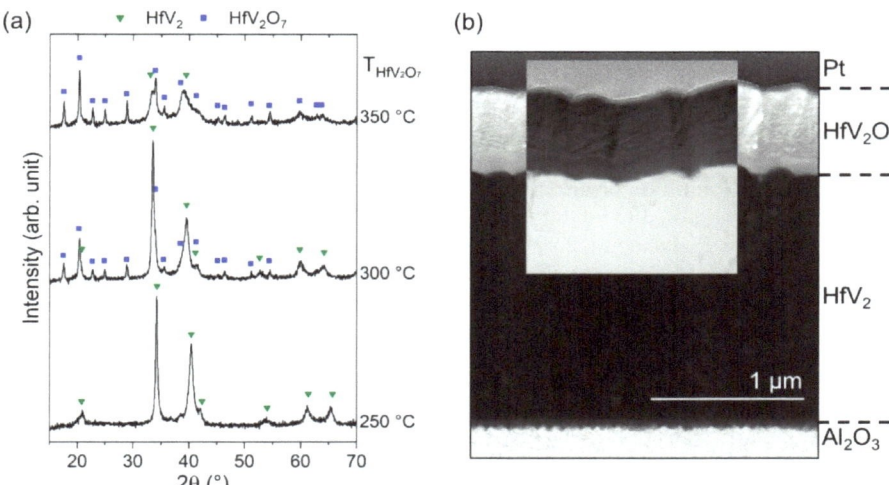

Figure 8. (**a**) Diffractograms of magnetron sputtered HfV$_2$–HfV$_2$O$_7$ bilayers. The synthesis temperature T for HfV$_2$O$_7$ on HfV$_2$ has been varied between 250 and 350 °C. (**b**) Scanning transmission electron microscopy bright-field with inset high-angle annular dark-field image of HfV$_2$–HfV$_2$O$_7$ cross-section with HfV$_2$O$_7$ having been synthesized at 300 °C.

For a synthesis temperature of 250 °C, peaks that are attributed to HfV$_2$ are barely shifted [78] indicating a low concentration of interstitially incorporated oxygen into the structure during the reactive sputtering process of HfV$_2$O$_7$. However, no distinct HfV$_2$O$_7$ peaks were obtained indicating an amorphous or nanocrystalline structure. It is expected that amorphous HfV$_2$O$_7$ does not exhibit NTE. While crystalline ZrW$_2$O$_8$ exhibits NTE in its entire stability range [38,39], PTE was obtained for amorphous ZrW$_2$O$_8$ films synthesized by reactive sputtering [42].

An increase in synthesis temperature to 300 °C results in more pronounced oxygen incorporation in HfV$_2$, as indicated by the increasing peak shift, which corresponds to the shift obtained for a 10-weeks exposure time of uncapped HfV$_2$ to air (see Figure 3). Besides, peaks of crystalline HfV$_2$O$_7$ emerge, demonstrating the first synthesis of a crystalline HfV$_2$–HfV$_2$O$_7$ composite. The effect of minute concentrations of interstitially solved oxygen on the thermoelastic anomaly of HfV$_2$ has not been studied yet and is hence proposed for future investigations. Figure 8b shows the corresponding film cross-section studied by STEM. The image suggests the formation of a bilayered structure with a defined interface separating the HfV$_2$ bottom layer (1.7 µm) from HfV$_2$O$_7$ (0.5 µm) top layer. Based on the atomic number (z) contrast using high-angle annular dark-field imaging, the distinct formation of HfV$_2$ (high z) and HfV$_2$O$_7$ (low z) is supported. The deposition rates for HfV$_2$ and HfV$_2$O$_7$ were determined to be approximately 22 and 5 nm min^{-1}, respectively. Thus, tuning the volume phase fractions to be in line with the proposed ones (see Figure 1) can simply be achieved by adjusting the deposition times for the individual layer. However, further investigations are required to evaluate whether these bilayered composite films allow for accurate determination of their integral temperature coefficient of elasticity and coefficient of thermal expansion. A further increase in the synthesis temperature of HfV_2O_7 to 350 °C reveals the onset of amorphization of HfV_2 due to oxygen incorporation, as indicated by the pronounced peak shifting and broadening.

4. Conclusions

The HfV_2–HfV_2O_7 composite has been proposed as a material with temperature-independent thermophysical properties due to the combination of the anomalously increasing thermoelastic constants of HfV_2 and the negative thermal expansion of HfV_2O_7. Both the temperature coefficient of elasticity and the coefficient of thermal expansion were predicted to be near zero for a phase fraction of approximately 30 vol.% HfV_2 and 70 vol.% HfV_2O_7.

Two synthesis pathways for HfV_2–HfV_2O_7 composites were studied: (1) annealing of magnetron sputtered HfV_2 thin films in air to form a HfV_2O_7 oxide scale on the thin film surface and (2) magnetron sputtering of HfV_2O_7/HfV_2 bilayers. The onset of the oxidation behavior of HfV_2 thin films is characterized by continuous interstitial incorporation of oxygen, occurring already at room temperature, finally yielding amorphization. Crystalline HfV_2O_7 forms at 550 °C. The high oxygen mobility in HfV_2 is suggested to inhibit the formation of crystalline HfV_2–HfV_2O_7 composites by annealing HfV_2 in air. Reducing the formation temperature of crystalline HfV_2O_7 down to 300 °C using reactive magnetron sputtering enables the synthesis of a crystalline bilayered HfV_2–HfV_2O_7 composite. The NTE properties of HfV_2O_7 were verified for monolithic magnetron sputtered HfV_2O_7 thin films using temperature-dependent in situ X-ray diffraction.

Author Contributions: P.K., A.L.R., and J.M.S. conceived and designed the experiments and DFT calculations. A.L.R. and P.K. performed the sputtering experiments, the EDX and XRD analyses, and the DFT calculations. D.M.H. supported A.L.R. with the temperature-dependent in situ XRD measurements. ERDA measurements and data analysis were conducted by M.H. and D.P. STEM was performed by M.H. and S.K.A. All authors contributed to the evaluation and interpretation of the experimental and theoretical results. The manuscript was primarily written by P.K. with input from all authors. All authors have read and agreed to the published version of the manuscript.

Funding: Simulations were performed with computing resources granted by JARA-HPC from RWTH Aachen University under project JARA0131. Support for the operation of the accelerator laboratory in Uppsala by VR-RFI (contract 2017-00646-9) and the Swedish Foundation for Strategic Research (SSF, contract RIF14-0053) is gratefully acknowledged. The authors also kindly acknowledge funding through the Max Planck Fellow Program.

Conflicts of Interest: The authors declare no conflict of interest.

References

1. Roy, R.; Agrawal, D.K.; McKinstry, H.A. Very low thermal expansion coefficient materials. *Annu. Rev. Mater. Sci.* **1989**, *19*, 59–81. [CrossRef]
2. Wolff, E.G. Thermal expansion in metal/lithia-alumina-silica (LAS) composites. *Int. J. Thermophys.* **1988**, *9*, 221–232. [CrossRef]
3. Mandal, S.; Chakrabarti, S.; Das, S.K.; Ghatak, S. Synthesis of low expansion ceramics in lithia–alumina–silica system with zirconia additive using the powder precursor in the form of hydroxyhydrogel. *Ceram. Int.* **2007**, *33*, 123–132. [CrossRef]
4. Lind, C. Two decades of negative thermal expansion research: Where do we stand? *Materials* **2012**, *5*, 1125–1154. [CrossRef]
5. Takenaka, K. Negative thermal expansion materials: Technological key for control of thermal expansion. *Sci. Technol. Adv. Mater.* **2012**, *13*, 013001. [CrossRef]
6. Feng Huang, L.; Zeng, Z. Lattice dynamics and disorder-induced contraction in functionalized graphene. *J. Appl. Phys.* **2013**, *113*, 083524. [CrossRef]

7. Saito, T.; Furuta, T.; Hwang, J.-H.; Kuramoto, S.; Nishino, K.; Suzuki, N.; Chen, R.; Yamada, A.; Ito, K.; Seno, Y.; et al. Multifunctional alloys obtained via a dislocation-free plastic deformation mechanism. *Science* **2003**, *300*, 464–467. [CrossRef]
8. Cahn, R.W. An unusual Nobel Prize. *Notes Rec. R. Soc.* **2005**, *59*, 145–153. [CrossRef]
9. Wang, Y.; Gao, J.; Wu, H.; Yang, S.; Ding, X.; Wang, D.; Ren, X.; Wang, Y.; Song, X.; Gao, J. Strain glass transition in a multifunctional β-type Ti alloy. *Sci. Rep.* **2014**, *4*, 3995. [CrossRef]
10. Oh, J.M.; Kang, J.-H.; Lee, S.; Kim, S.-D.; Kang, N.; Park, C.H. Origin of superproperties of Ti-23Nb-1Ta-2Hf-O alloy. *Mater. Lett.* **2018**, *233*, 162–165. [CrossRef]
11. Qin, F.; Lu, F.; Chen, Y.; Yang, J.; Zhao, X. Deformation induced elinvar behavior in Fe–Ni invar alloy. *Sci. Bull.* **2018**, *63*, 1040–1042. [CrossRef]
12. Dove, M.T.; Fang, H. Negative thermal expansion and associated anomalous physical properties: Review of the lattice dynamics theoretical foundation. *Rep. Prog. Phys.* **2016**, *79*, 066503. [CrossRef] [PubMed]
13. Grimvall, G. *Thermophysical Properties of Materials*; Elsevier Science B.V.: Amsterdam, The Netherlands, 1999. [CrossRef]
14. Wachtman, J.B.; Tefft, W.E.; Lam, D.G.; Apstein, C.S. Exponential temperature dependence of young's modulus for several oxides. *Phys. Rev.* **1961**, *122*, 1754–1759. [CrossRef]
15. Chen, J.; Nittala, K.; Forrester, J.S.; Jones, J.L.; Deng, J.; Yu, R.; Xing, X. The role of spontaneous polarization in the negative thermal expansion of tetragonal $PbTiO_3$-based compounds. *J. Am. Chem. Soc.* **2011**, *133*, 11114–11117. [CrossRef]
16. Yamada, I.; Tsuchida, K.; Ohgushi, K.; Hayashi, N.; Kim, J.; Tsuji, N.; Takahashi, R.; Matsushita, M.; Nishiyama, N.; Inoue, T.; et al. Giant negative thermal expansion in the iron perovskite $SrCu_3Fe_4O_{12}$. *Angew. Chem. Int. Ed.* **2011**, *50*, 6579–6582. [CrossRef]
17. Romao, C.P.; Miller, K.J.; Whitman, C.A.; White, M.A. *Comprehensive Inorganic Chemistry II Negative Thermal Expansion (Thermomiotic) Materials*; Elsevier: Oxford, UK, 2013; Volume 4, pp. 128–151.
18. Takenaka, K.; Takagi, H. Giant negative thermal expansion in Ge-doped anti-perovskite manganese nitrides. *Appl. Phys. Lett.* **2005**, *87*, 261902. [CrossRef]
19. Walker, E. Anomalous temperature behaviour of the shear elastic constant C_{44} in vanadium. *Solid State Commun.* **1978**, *28*, 587–589. [CrossRef]
20. Armstrong, P.E.; Dickinson, J.M.; Brown, H.L. Temperature dependence of the elastic stiffness coefficients of niobium (columbium). *Trans. Am. Inst. Min. Metall. Pet. Eng.* **1966**, *236*, 1404.
21. Walker, E.; Bujard, P. Anomalous temperature behaviour of the shear elastic constant C_{44} in tantalum. *Solid State Commun.* **1980**, *34*, 691–693. [CrossRef]
22. Rayne, J.A. Elastic Constants of Palladium from 4.2-300 K. *Phys. Rev.* **1960**, *118*, 1545–1549. [CrossRef]
23. Macfarlane, R.E.; Rayne, J.A.; Jones, C.K. Anomalous temperature dependence of shear modulus C_{44} for Platinum. *Phys. Lett.* **1965**, *18*, 91–92. [CrossRef]
24. Walker, E.; Ortelli, J.; Peter, M. Elastic constants of monocrystalline alloys of Pd–Rh and Pd–Ag between 4.2 K and 300 K. *Phys. Lett. A* **1970**, *31*, 240–241. [CrossRef]
25. Ashkenazi, J.; Dacorogna, M.; Peter, M.; Talmor, Y.; Walker, E.; Steinemann, S. Elastic constants in Nb–Zr alloys from zero temperature to the melting point: Experiment and theory. *Phys. Rev. B* **1978**, *18*, 4120–4131. [CrossRef]
26. Bujard, P.; Sanjines, R.; Walker, E.; Ashkenazi, J.; Peter, M. Elastic constants in Nb–Mo alloys from zero temperature to the melting point: Experiment and theory. *J. Phys. F* **1981**, *11*, 775. [CrossRef]
27. Keuter, P.; Music, D.; Schnabel, V.; Stuer, M.; Schneider, J.M. From qualitative to quantitative description of the anomalous thermoelastic behavior of V, Nb, Ta, Pd and Pt. *J. Phys. Condens. Matter* **2019**, *31*, 225402. [CrossRef]
28. Keuter, P.; Music, D.; Stuer, M.; Schneider, J.M. Electronic structure tuning of the anomalous thermoelastic behavior in Nb–X (X = Zr, V, Mo) solid solutions. *J. Appl. Phys.* **2019**, *125*, 215103. [CrossRef]
29. Keuter, P.; Music, D.; Stuer, M.; Schneider, J.M. Temperature and impurity induced stabilization of cubic HfV_2 laves phase. *Condens. Matter* **2019**, *4*, 63. [CrossRef]
30. Huang, L.; Vitos, L.; Kwon, S.K.; Johansson, B.; Ahuja, R. Thermoelastic properties of random alloys from first-principles theory. *Phys. Rev. B* **2006**, *73*, 104203. [CrossRef]
31. Huang, L.; Ramzan, M.; Vitos, L.; Johansson, B.; Ahuja, R. Anomalous temperature dependence of elastic constant c_{44} in V, Nb, Ta, Pd, and Pt. *J. Phys. Chem. Solids* **2010**, *71*, 1065–1068. [CrossRef]

32. Lüthi, B.; Herrmann, M.; Assmus, W.; Schmidt, H.; Rietschel, H.; Wühl, H.; Gottwick, U.; Sparn, G.; Steglich, F. Normal-state and superconducting properties of HfV_2. *Z. Phys. B Condens. Matter* **1985**, *60*, 387–392. [CrossRef]
33. Rudy, E.; Windisch, S. The phase diagrams hafnium-vanadium and hafnium-chromium. *J. Less Common. Met.* **1968**, *15*, 13–27. [CrossRef]
34. Finlayson, T.R.; Lanston, E.J.; Simpson, M.A.; Gibbs, E.E.; Smith, T.F. Elastic properties of $(Hf,Zr)V_2$ superconducting compounds. *J. Phys. F* **1978**, *8*, 2269. [CrossRef]
35. Balankin, A.S.; Skorov, D.M. Anomalies of elastic moduli in ZrV_2 and HfV_2 Laves phases at high temperatures. *Sov. Phys. Solid State* **1982**, *24*, 681–682.
36. Pushkarev, E.A.; Petrenko, N.S.; Finkel, V.A. Thermal expansion of the superconducting compound HfV_2 at low temperatures. *Phys. Status Solidi A* **1978**, *47*, K145–K148. [CrossRef]
37. Attfield, J.P. Mechanisms and Materials for NTE. *Front. Chem.* **2018**, *6*. [CrossRef] [PubMed]
38. Mary, T.A.; Evans, J.S.O.; Vogt, T.; Sleight, A.W. Negative thermal expansion from 0.3 to 1050 Kelvin in ZrW_2O_8. *Science* **1996**, *272*, 90–92. [CrossRef]
39. Evans, J.S.O.; Mary, T.A.; Sleight, A.W. Negative thermal expansion materials. *Phys. B Condens. Matter* **1997**, *241–243*, 311–316. [CrossRef]
40. Hisashige, T.; Yamaguchi, T.; Tsuji, Y.; Yamamura, T. Phase Transition of $Zr_{1-x}Hf_xV_2O_7$ solid solutions having negative thermal expansion. *J. Ceram. Soc. Jpn.* **2006**, *114*, 607–611. [CrossRef]
41. Turquat, C.; Muller, C.; Nigrelli, E.; Leroux, C.; Soubeyroux, J.L.; Nihoul, G. Structural investigation of temperature-induced phase transitions in HfV_2O_7. *Eur. Phys. J. Appl. Phys.* **2000**, *10*, 15–27. [CrossRef]
42. Sutton, M.S.; Talghader, J. Micromachined negative thermal expansion thin films. *J. Microelectromechanical Syst.* **2004**, *13*, 1148–1151. [CrossRef]
43. Liu, H.; Yang, L.; Zhang, Z.; Pan, K.; Zhang, F.; Cheng, H.; Zeng, X.; Chen, X. Preparation and optical, nanomechanical, negative thermal expansion properties of $Sc_2W_3O_{12}$ thin film grown by pulsed laser deposition. *Ceram. Int.* **2016**, *42*, 8809–8814. [CrossRef]
44. Liu, H.; Zhang, Z.; Zhang, W.; Chen, X.; Cheng, X. Negative thermal expansion ZrW_2O_8 thin films prepared by pulsed laser deposition. *Surf. Coat. Technol.* **2011**, *205*, 5073–5076. [CrossRef]
45. Liu, H.; Pan, K.; Jin, Q.; Zhang, Z.; Wang, G.; Zeng, X. Negative thermal expansion and shift in phase transition temperature in Mo-substituted ZrW_2O_8 thin films prepared by pulsed laser deposition. *Ceram. Int.* **2014**, *40*, 3873–3878. [CrossRef]
46. Bespalov, I.; Datler, M.; Buhr, S.; Drachsel, W.; Rupprechter, G.; Suchorski, Y. Initial stages of oxide formation on the Zr surface at low oxygenpressure: An in situ FIM and XPS study. *Ultramicroscopy* **2015**, *159*, 147–151. [CrossRef]
47. Laguna, O.H.; Pérez, A.; Centeno, M.A.; Odriozola, J.A. Synergy between gold and oxygen vacancies in gold supported on Zr-doped ceria catalysts for the CO oxidation. *Appl. Catal. B* **2015**, *176–177*, 385–395. [CrossRef]
48. Schneider, J.M.; Hjörvarsson, B.; Wang, X.; Hultman, L. On the effect of hydrogen incorporation in strontium titanate layers grown by high vacuum magnetron sputtering. *Appl. Phys. Lett.* **1999**, *75*, 3476–3478. [CrossRef]
49. Ravensburg, A.L.; Keuter, P.; Music, D.; Miljanovic, D.J.; Schneider, J.M. Experimental and Theoretical Investigation of the Elastic Properties of HfV_2O_7. *Crystals* **2020**, *10*, 172. [CrossRef]
50. Zhang, Y.; Whitlow, H.J.; Winzell, T.; Bubb, I.F.; Sajavaara, T.; Arstila, K.; Keinonen, J. Detection efficiency of time-of-flight energy elastic recoil detection analysis systems. *Nucl. Instrum. Methods Phys. Res. B* **1999**, *149*, 477–489. [CrossRef]
51. Janson, M.S. *Contes Instruction Manual*; Uppsala University: Uppsala, Sweden, 2004.
52. Bragg, W.H.; Bragg, W.L. The reflection of X-rays by crystals. *Proc R. Soc. Lond. Ser. A Contain. Pap. Math. Phys. Character* **1913**, *88*, 428–438. [CrossRef]
53. Langford, R.M.; Rogers, M. In situ lift-out: Steps to improve yield and a comparison with other FIB TEM sample preparation techniques. *Micron* **2008**, *39*, 1325–1330. [CrossRef]
54. Hohenberg, P.; Kohn, W. Inhomogeneous electron gas. *Phys. Rev.* **1964**, *136*, B864–B871. [CrossRef]
55. Kresse, G.; Hafner, J. Ab initio molecular dynamics for open-shell transition metals. *Phys. Rev. B* **1993**, *48*, 13115–13118. [CrossRef] [PubMed]

56. Kresse, G.; Hafner, J. Ab initio molecular-dynamics simulation of the liquid-metal-amorphous-semiconductor transition in germanium. *Phys. Rev. B* **1994**, *49*, 14251–14269. [CrossRef] [PubMed]
57. Perdew, J.P.; Burke, K.; Ernzerhof, M. Generalized gradient approximation made simple. *Phys. Rev. Lett.* **1996**, *77*, 3865–3868. [CrossRef]
58. Monkhorst, H.J.; Pack, J.D. Special points for Brillouin-zone integrations. *Phys. Rev. B* **1976**, *13*, 5188–5192. [CrossRef]
59. Birch, F. Finite Eastic Strain of Cubic Crystals. *Phys. Rev.* **1947**, *71*, 809–824. [CrossRef]
60. Murnaghan, F.D. The compressibility of media under extreme pressures. *Proc. Natl. Acad. Sci. USA* **1944**, *15*, 244–247. [CrossRef]
61. Drymiotis, F.R.; Ledbetter, H.; Betts, J.B.; Kimura, T.; Lashley, J.C.; Migliori, A.; Ramirez, A.P.; Kowach, G.R.; Van Duijn, J. Monocrystal elastic constants of the negative-thermal-expansion compound zirconium tungstate (ZrW_2O_8). *Phys. Rev. Lett.* **2004**, *93*, 025502. [CrossRef]
62. De Buysser, K.; Lommens, P.; De Meyer, C.; Bruneel, E.; Hoste, S.; Van Driessche, I. ZrO_2-ZrW_2O_8 composites with tailor-made thermal expansion. *Ceram. Silik.* **2004**, *48*, 139–144.
63. Evans, J.S.O.; Mary, T.A.; Vogt, T.; Subramanian, M.A.; Sleight, A.W. Negative thermal expansion in ZrW_2O_8 and HfW_2O_8. *Chem. Mater.* **1996**, *8*, 2809–2823. [CrossRef]
64. Yamamura, Y.; Horikoshi, A.; Yasuzuka, S.; Saitoh, H.; Saito, K. Negative thermal expansion emerging upon structural phase transition in ZrV_2O_7 and HfV_2O_7. *Dalt. Trans.* **2011**, *40*, 2242–2248. [CrossRef] [PubMed]
65. Romao, C.P.; Marinkovic, B.A.; Werner-Zwanziger, U.; White, M.A. Thermal expansion reduction in alumina-toughened zirconia by incorporation of zirconium tungstate and aluminum tungstate. *J. Am. Ceram. Soc.* **2015**, *98*, 2858–2865. [CrossRef]
66. Kingery, W.D.; Bowen, H.K.; Uhlmann, D.R. *Introduction to Ceramics*; Wiley: New York, NY, USA, 1960.
67. Chen, H.-Y.; Lu, F.-H. Oxidation behavir of titanium nitride films. *J. Vac. Sci. Technol. A* **2005**, *23*. [CrossRef]
68. Stelzer, B.; Momma, M.; Schneider, J.M. Autonomously Self-Reporting Hard Coatings: Tracking the temporal oxidation behavior of tin by in situ sheet resistance measurements. *Adv. Funct. Mater.* **2020**, *30*, 2000146. [CrossRef]
69. Eickert, S.; Hecht, H.; von Minnigerode, G. Formation area of amorphous thin V–Zr films prepared by cocondensation on hot substrates. *Z. Phys. B Condens. Matter* **1992**, *88*, 35–38. [CrossRef]
70. Shi, L.Q.; Xu, S.L. Phase transformations of sputtered ZrV_2 films after annealing and hydrogenation. *J. Vac. Sci. Technol. A* **2006**, *24*, 190–194. [CrossRef]
71. King, D.J.M.; Middleburgh, S.C.; Liu, A.C.Y.; Tahini, H.A.; Lumpkin, G.R.; Cortie, M.B. Formation and structure of V–Zr amorphous alloy thin films. *Acta Mater.* **2015**, *83*, 269–275. [CrossRef]
72. Chihi, T.; Fatmi, M.; Ghebouli, B. Ab initio calculations for properties of laves phase V_2M (M = Zr, Hf, Ta) compounds. *Am. J. Mod. Phys.* **2013**, *2*, 88–92. [CrossRef]
73. Lumley, S.C.; Murphy, S.T.; Burr, P.A.; Grimes, R.W.; Chard-Tuckey, P.R.; Wenman, M.R. The stability of alloying additions in Zirconium. *J. Nucl. Mater.* **2013**, *437*, 122–129. [CrossRef]
74. Levy, O.; Hart, G.L.W.; Curtarolo, S. Hafnium binary alloys from experiments and first principles. *Acta Mater.* **2010**, *58*, 2887–2897. [CrossRef]
75. Vřešťál, J.; Pavlů, J.; Wdowik, U.D.; Šob, M. Modelling of phase equilibria in the Hf-V system below room temperature. *J. Min. Metall. Sect. B Metall.* **2017**, *53*, 239–247. [CrossRef]
76. Parsons, M.J.; Brown, P.J.; Crangle, J.; Neumann, K.U.; Ouladdiaf, B.; Smith, T.J.; Zayer, N.K.; Ziebeck, K.R.A. A study of the structural phase transformation and superconductivity in HfV_2. *J. Phys. Condens. Matter* **1998**, *10*, 8523. [CrossRef]
77. Zhao, Y.; Chu, F.; Von Dreele, R.B.; Zhu, Q. Structural phase transitions of HfV_2 at low temperatures. *Acta Crystallogr. Sect. B Struct. Sci.* **2000**, *56*, 601–606. [CrossRef]
78. Rapp, Ö.; Benediktsson, G. Latent heat of structural transformations in ZrV_2 and HfV_2. *Phys. Lett. A* **1979**, *74*, 449–452. [CrossRef]
79. Kim, W.-Y.; Luzzi, D.E.; Pope, D.P. Room temperature deformation behavior of the Hf–V–Ta C15 Laves phase. *Intermetallics* **2003**, *11*, 257–267. [CrossRef]
80. Bolvardi, H.; Emmerlich, J.; Mráz, S.; Arndt, M.; Rudigier, H.; Schneider, J.M. Low temperature synthesis of Mo_2BC thin films. *Thin Solid Films* **2013**, *542*, 5–7. [CrossRef]
81. Smith, T.F.; Shelton, R.N.; Lawson, A.C. Superconductivity and structural instability of (Hf, Zr)V_2 and (Hf, Ta)V_2 alloys at high pressure. *J. Phys. F* **1973**, *3*, 2157. [CrossRef]

82. Forker, M.; Herz, W.; Simon, D. Impurity trapping in the laves phase HfV$_2$ detected by perturbed angular correlations. *J. Phys. Condens. Matter* **1992**, *4*, 213. [CrossRef]
83. Greczynski, G.; Petrov, I.; Greene, J.E.; Hultman, L. Al capping layers for nondestructive x-ray photoelectron spectroscopy analyses of transition-metal nitride thin films. *J. Vac. Sci. Technol. A* **2015**, *33*, 05E101. [CrossRef]
84. Cramer, S.D.; Covino, B.S. *ASM Metals Handbook—Corrosion: Fundamentals, Testing, and Protection*; ASM International: Materials Park Campus, OH, USA, 2003; Volume 13A.
85. Kang, Y.-B. Critical evaluation and thermodynamic optimization of the VO–VO$_{2.5}$ system. *J. Eur. Ceram. Soc.* **2012**, *32*, 3187–3198. [CrossRef]
86. Wang, J.; Li, H.P.; Stevens, R. Hafnia and hafnia-toughened ceramics. *J. Mater. Sci.* **1992**, *27*, 5397–5430. [CrossRef]
87. Turquat, C.; Leroux, C.; Gloter, A.; Serin, V.; Nihoul, G. V-doped HfO$_2$: Thermal stability and vanadium valence. *Int. J. Inorg. Mater.* **2001**, *3*, 1025–1032. [CrossRef]
88. Turquat, C.; Leroux, C.; Roubin, M.; Nihoul, G. Vanadium-doped hafnia: Elaboration and structural characterization. *Solid State Sci.* **1999**, *1*, 3–13. [CrossRef]
89. Manory, R.; Mori, T.; Shimizu, I.; Miyake, S.; Kimmel, G. Growth and structure control of HfO$_{2-x}$ films with cubic and tetragonal strucutres obtained by ion beam assisted deposition. *J. Vac. Sci. Technol. A* **2002**, *20*. [CrossRef]
90. Hemamala, U.L.C.; El-Ghussein, F.; Goedken, A.M.; Chen, B.; Leroux, C.; Kruger, M.B. High-pressure x-ray diffraction and Raman spectroscopy of HfV$_2$O$_7$. *Phys. Rev. B* **2004**, *70*, 70. [CrossRef]
91. Pryde, A.K.A.; Hammonds, K.D.; Dove, M.T.; Heine, V.; Gale, J.D.; Warren, M.C. Origin of the negative thermal expansion in ZrW$_2$O$_8$ and ZrV$_2$O$_7$. *J. Phys. Condens. Matter* **1996**, *8*, 10973–10982. [CrossRef]
92. Mittal, R.; Chaplot, S.L. Lattice dynamica calculation of negative thermal expansion in ZrV$_2$O$_7$ and HfV$_2$O$_7$. *Phys. Rev. B* **2008**, *78*. [CrossRef]
93. Korthuis, V.; Khosrovani, N.; Sleight, A.W.; Roberts, N.; Dupree, R.; Warren, W.W., Jr. Negative Thermal expansion and phase transitions in the ZrV$_{2-x}$P$_x$O$_7$ series. *Chem. Mater.* **1995**, *7*, 412–417. [CrossRef]
94. Evans, J.S.O.; Hanson, J.C.; Sleight, A.W. Room-temperature superstructure of ZrV$_2$O$_7$. *Acta Crystallogr. Sect. B* **1998**, *54*, 705–713. [CrossRef]
95. Khosrovani, N.; Sleight, A.W.; Vogt, T. Structure of ZrV$_2$O$_7$ from −263 to 470 °C. *J. Solid State Chem.* **1997**, *132*, 355–360. [CrossRef]
96. Withers, R.L.; Evans, J.S.O.; Hanson, J.; Sleight, A.W. An in situ temperature-dependent electron and X-ray diffraction study of structural phase transitions in ZrV$_2$O$_7$. *J. Solid State Chem.* **1998**, *137*, 161–167. [CrossRef]
97. Baran, E.J. The unit cell of hafnium divanadate. *J. Less Common. Met.* **1976**, *46*, 343–345. [CrossRef]

Publisher's Note: MDPI stays neutral with regard to jurisdictional claims in published maps and institutional affiliations.

© 2020 by the authors. Licensee MDPI, Basel, Switzerland. This article is an open access article distributed under the terms and conditions of the Creative Commons Attribution (CC BY) license (http://creativecommons.org/licenses/by/4.0/).

Communication

A Novel Microstructural Evolution Model for Growth of Ultra-Fine Al₂O₃ Oxides from SiO₂ Silica Ceramic Decomposition during Self-Propagated High-Temperature Synthesis

Mateusz Kopec [1,*], Stanisław Jóźwiak [2] and Zbigniew L. Kowalewski [1]

1. Institute of Fundamental Technological Research, Polish Academy of Sciences, Pawińskiego 5B, 02-106 Warszawa, Poland; zkowalew@ippt.pan.pl
2. Faculty of Advanced Technologies and Chemistry, Military University of Technology, 00-908 Warsaw 49, Poland; Stanislaw.jozwiak@wat.edu.pl
* Correspondence: mkopec@ippt.pan.pl

Received: 27 May 2020; Accepted: 16 June 2020; Published: 23 June 2020

Abstract: In this paper, experimental verification of the microstructural evolution model during sintering of aluminum, iron and particulate mullite ceramic powders using self-propagated high-temperature synthesis (SHS) was performed. The powder mixture with 20% wt. content of reinforcing ceramic was investigated throughout this research. The mixed powders were cold pressed and sintered in a vacuum at 1030 °C. The SHS reaction between sintered feed powders resulted in a rapid temperature increase from the heat generated. The temperature increase led to the melting of an aluminum-based metallic liquid. The metallic liquid infiltrated the porous SiO_2 ceramics. Silicon atoms were transited into the intermetallic iron–aluminum matrix. Subsequently, a ternary matrix from the Fe–Al–Si system was formed, and synthesis of the oxygen and aluminum occurred. Synthesis of both these elements resulted in formation of new, fine Al_2O_3 precipitates in the volume of matrix. The proposed microstructural evolution model for growth of ultra-fine Al_2O_3 oxides from SiO_2 silica ceramic decomposition during SHS was successfully verified through scanning electron microscopy (SEM), X-ray energy-dispersive spectroscopy (EDS) analysis and X-ray diffraction (XRD).

Keywords: intermetallics; powder methods; electron microscopy; X-ray analysis

1. Introduction

Intermetallic–ceramic composites (IMCs) are a narrow group of composites used as structural and functional materials. These composites combine unique properties of ceramics (hardness, thermo-chemical resistance and thermal stability) with properties of metals, i.e., mechanical strength. Due to their superior properties, there is a wide spectrum of possible applications for these materials [1]. Composites, in comparison to conventional structural materials, are characterized by a higher Young's modulus, a low coefficient of thermal expansion, abrasion resistance and high strength. These properties could be also maintained during operation at elevated temperatures [2].

The potential applications of IMC materials have led to a growing number of publications describing their properties and manufacturing technologies. A number of studies investigating the manufacturing processes of IMCs reinforced by iron, aluminum and oxide particles with various mass contents of strengthening elements—Fe (15%–93%), Al (5%–70%) and Si (0.5%–25%)—are found in the literature [3–8].

The review of IMC manufacturing techniques presents a wide spectrum of possibilities for IMC fabrication from conventional sintering methods [4] to novel and advanced techniques: for example, diffusion bonding of plates made of high-purity input materials [5–7]. For commercial applications, it is possible to obtain the final product by using a conventional sintering method, reactive sintering [9] or advanced self-propagated high-temperature synthesis (SHS) [3,4,10]. An important aspect in manufacturing high-performance IMCs is the use of the finest possible reinforcing phase powders (<1 μm) during the sintering process. By using such powders, a relatively good distribution of reinforcement in the matrix volume and its location in the sinter volume can be achieved. Fine dimensions of reinforcement powders might be obtained by using high-energy milling techniques [11–15]. Satisfactory effects of powder fragmentation have been observed after a long milling time of over 5 h [16,17]. A long milling time includes technological breaks for stabilization of milling conditions and results in a long processing time and low efficiency of the whole process. In the wide spectrum of IMCs, Al_2O_3 reinforced composites are characterized by the best mechanical properties [18–20]. They are usually reinforced with dispersive Al_2O_3 oxides obtained during synthesis or thermite reaction [20–27].

Different approaches to model the sintering processes and microstructure evolutions are reported in the literature. These include phenomenological and mechanistic methods, continuous and discrete formulations, modeling at both macro and micro levels, as well as multiscale modeling [28–32]. Other methods include microstructural evolution models. An interesting approach reported in [33] allows for development of a phase-field model to investigate intermetallic compound phase transformation and nucleation of one of the phases involved. On the one hand, phase-field methods are considered as powerful tools for microstructure modeling [34]. They could be used to model various processes, including rapid solidification [35], additive manufacturing [36] and various materials [37,38]. On the other hand, new studies on microstructural evolution during heating of carbon nanotube metal matrix composites (CNT-MMC) with additional modeling attempts have been recently reported [39]. The variety of materials and composites extends the possibilities of microstructure evolution modeling to different methods and approaches.

Based on the detailed literature review, a novel microstructure evolution model for investigating powder mixtures under high-temperature sintering was developed. The rapid temperature increase, generated during SHS, led to the melting of aluminum-based metallic liquid. The metallic liquid infiltrated the porous SiO_2 ceramics, and silicon atoms were transited into the intermetallic iron–aluminum matrix. Subsequently, formation of an Fe–Al–Si ternary matrix and synthesis of the oxygen and aluminum occurred. Synthesis of both these elements resulted in formation of new, fine Al_2O_3 precipitates in the volume of matrix. A model for growth of ultra-fine Al_2O_3 oxides from decomposition of SiO_2 silica during self-propagated high-temperature synthesis was experimentally verified through vacuum sintering, combined with high-energy milling of reinforcement particles. This method allowed for obtainment of a permanent connection between matrix and reinforcement in the fabricated IMC/Al_2O_3 composite. Additionally, high-energy powder milling of the reinforcing phase promises a reduction of the composite inhomogeneity after high-temperature sintering.

2. Materials and Methods

The powder mixture of 28% wt. of iron and 52 wt.% of aluminum with 20 wt.% of the reinforcing mullite ceramic was proposed to evaluate the microstructure evolution model. The specific contents of aluminum and iron were selected to obtain an intermetallic matrix of the sintered material. Milling parameters were selected on the basis of literature analysis, the high-energy milling machine manufacturer's recommendations and the authors' experience. A grinding jar and milling balls of 10 mm diameter were made of 100 Cr6 steel. The milling balls mass to feed powder mass ratio was equal to 10:1. The powder mixture was grinded for 5 min with a rotating speed of 450 RPM. After initial grinding, the mixture of mullite, iron and aluminum powders was put into a turbulent mixer for 30 min. The mixed powders were then pressed in a 25-mm-diameter die in a single-sided hydraulic press at 900 MPa for 3 min. The compacts obtained were located inside the graphite dies and sintered in a vacuum chamber at 1030 °C for 5 min without external load. The microstructural characterization was performed on an FEI Scios field emission gun scanning electron microscope (FEG-SEM) operated at 20 kV. Prior to this study, the specimens were first hot mounted and then ground using 80, 180, 300, 600, 800, 1200 and 4000 SiC paper. The polishing was performed using Metrep® MD-Chem cloth with 3 µm diamond suspension. The XRD measurements were performed using a Rigaku (Tokyo, Japan) Ultima IV diffractometer with Co-K radiation ($\lambda \frac{1}{4} 1.78897$ Å) and operating parameters of 40 mA and 40 kV with a scanning speed of 1°/min and a scanning step of 0.02° in the range of 20°–120°.

3. Introduction of Microstructure Evolution Model for Growth of Ultra-Fine Al_2O_3 Oxides from SiO_2 Silica Ceramic Decomposition

Literature analysis allowed for proposal of a model for microstructural evolution during high-temperature sintering of IMCs by using powder metallurgy technology. This model includes the sintering process of iron, aluminum and mullite ceramics, and simultaneous formation of oxide precipitates formed from the defragmentation of SiO_2 during the infiltration process. The model is mainly based on the SHS reaction between sintered feed powders (Figure 1a), resulting in a rapid temperature increase from the heat generated during the SHS. The temperature increase led to the melting of aluminum-based metallic liquid. The metallic liquid infiltrated the porous ceramics, which led to their defragmentation (Figure 1b). The mechanism of the oxide precipitates formation on the basis of SiO_2 decomposition can be represented by the following relationship:

$$SiO_2 + Fe_xAl_x \rightarrow Fe_xAl_xSi_x + Al_2O_3 \tag{1}$$

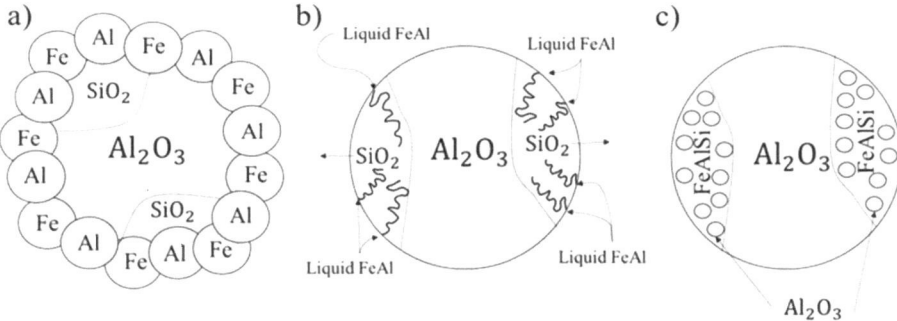

Figure 1. Schematic model of microstructure evolution during the self-propagated high-temperature synthesis (SHS) reaction: (**a**) initiation of the SHS reaction between sintered feed powders, (**b**) defragmentation of porous ceramics through infiltration, and (**c**) formation of new, fine Al_2O_3 precipitates in the volume of matrix.

Mullite ceramics exhibit wettability of liquid aluminum at 1000 °C [40], and thus an infiltration process may accelerate their defragmentation. During the sintering process, silicon atoms are expected to be transited into the intermetallic iron–aluminum matrix. Such a transition leads to the formation of a ternary matrix from the Fe–Al–Si system and subsequent synthesis of the oxygen and aluminum. Synthesis of both these elements results in formation of new, fine Al_2O_3 precipitates in the volume of matrix (Figure 1c) within the primary areas of silica ceramics. Finally, an IMC composite with a permanent connection between the ceramics and the intermetallic matrix can be obtained.

4. Results and Discussion

4.1. Optimization of Sintering Temperature for Investigated Powder Mixtures

The initial sintering parameters were selected for the temperature of 1200 °C, compressing load of 25 kN and sintering time of 5 min. However, these parameters did not allow for obtainment of a homogeneous structure of sintered material. During the sintering process, the pressure inside the graphite dies was so high that it caused them to crack. Therefore, in subsequent stages, sintering without external load was proposed. Unfortunately, during the process, liquid matrix material leaked between the graphite die and the upper punch pressing the sintered compact. Leaked material was characterized by using scanning electron microscopy (SEM) and chemical composition analysis (EDS), as presented in Figure 2 and Table 1. It was found that the contents of silicon and aluminum within the microstructure of material tested (Figure 2) were comparable to the hypoeutectic silumin characterized by the melting point of 577 °C. In order to prevent material loss during the heating stage, the sintering parameters were modified accordingly. It was found that an absence of metallic liquid leaks was obtained during sintering without external load and subsequent reduction of the sintering temperature. The sintering temperature was reduced to 1030 °C. Relatively small but obvious differences in the sintering temperature were caused by a change in the silicon content in the material structure. The higher content of silicon corresponded to the lower value of melting temperature. An application of new sintering parameters allowed for fabrication of the sinters, without any material losses, and moreover for changes in the chemical composition of the material.

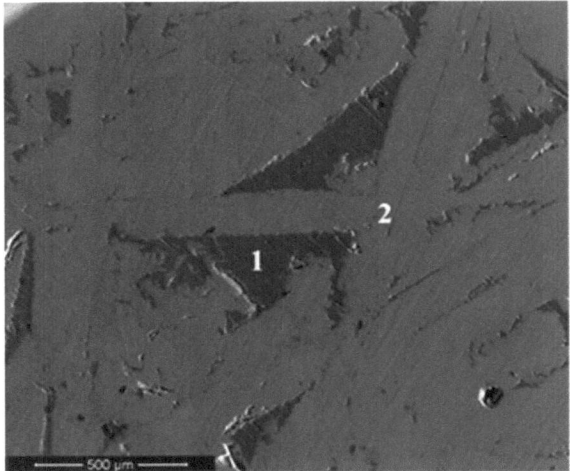

Figure 2. Microstructure of material leaked during sintering at 1200 °C.

Table 1. Chemical composition of material leaked during sintering at 1200 °C.

Wt.%	Al	Si	Fe
1	87.95	8.92	3.13
2	76.18	1.91	21.94

In order to obtain a homogeneous phase structure of the matrix, it was necessary to perform a homogenization process. It is widely accepted that in order to have more effective homogenization, a higher temperature in the process should be used. Higher values of temperature accelerate atom diffusion and subsequent phase transformations. The temperature was selected based on the analysis of the ternary aluminum–iron–silicon equilibrium system [41]. The temperature of 900 °C was selected in order to avoid the melting point identified on the basis of the chemical composition analysis of the phases formed after the sintering process. The sintering time of 5 min was used on the basis of the authors' research [42], where the optimized structure and properties of FeAl intermetallic phases were obtained.

4.2. Microstructural Characterization of Sintered Material and Experimental Verification of the Model

A microstructure of the sintered sample with 20% wt. reinforcement was characterized by the large, nonfragmented areas of the primary oxide precipitates in a ternary iron–aluminum–silicon matrix (Figure 3). The oxide ceramics were evenly distributed in the volume of the material, as presented in Figure 3a. The material matrix consisted of the $FeAl_3$ and $Al_{4.5}FeSi$ (τ_6) phases (Figure 3b). The τ_6 phase was mainly observed in the areas close to the ceramic reinforcement.

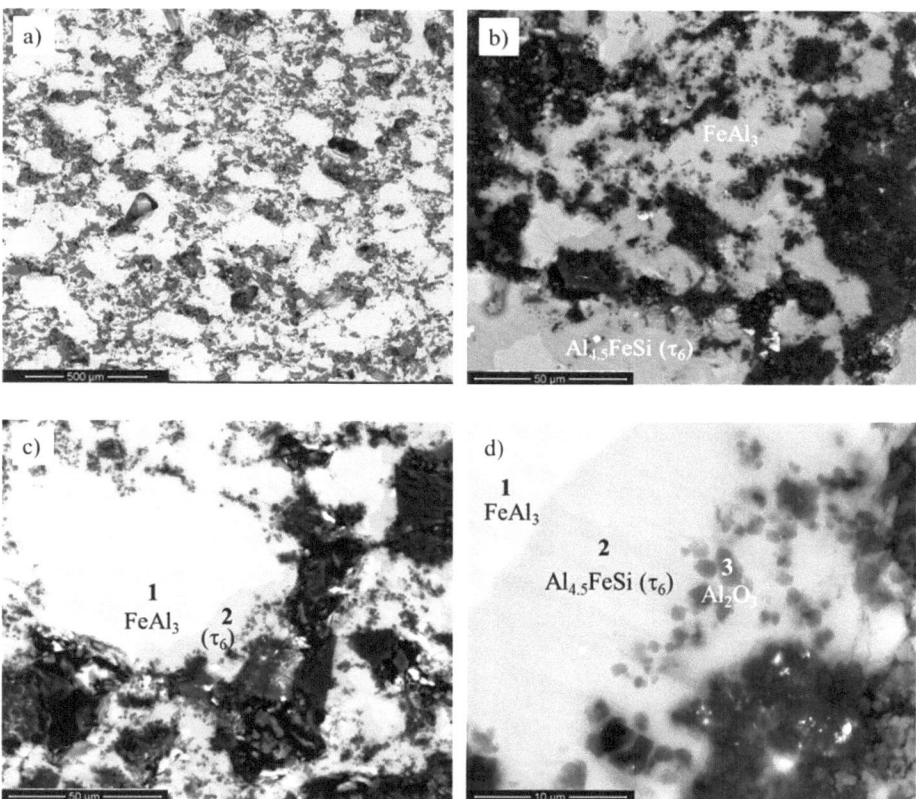

Figure 3. Microstructure of composite with 20% wt. content of mullite ceramics after sintering at 1030 °C for 5 min (**a**); $FeAl_3$ + $Al_{4.5}FeSi$ (τ_6) multiphase matrix (**b**); an initiation of the disintegration process of primary SiO_2 precipitates (**c**); silicon diffusion into the intermetallic iron–aluminum matrix and newly formed Al_2O_3 ceramics in the areas of primary SiO_2 particles (**d**).

The correctness of the model assumed was demonstrated by microanalysis of the chemical composition performed in the area of primary defragmented SiO_2 porous ceramic particles (Figure 3c,d and Table 2). It was observed that the dark areas ("3") were characterized with an increased proportion of oxygen and aluminum in comparison to the light areas ("1", "2"), indicating the dominant content of alundum ceramics. The light gray areas were characterized by increased contents of silicon and iron. The contents of silicon and iron in the material matrix confirmed that these specific zones were formed by the transition of silicon atoms from SiO_2 ceramics to the intermetallic matrix.

Table 2. Chemical composition of the decomposed area.

Wt.%	Al	Si	Fe	O
1	62.39	2.93	34.68	-
2	55.69	14.98	29.33	-
3	61.27	-	8.73	30.00

According to the microstructure evolution model and microstructural observations, phase transformations were described by the following equation:

$$FeAl_3 + SiO_2 \rightarrow Al_{4.5}FeSi\ (\tau_6) + Al_2O_3 \qquad (2)$$

One can indicate that high-energy milling led to defragmentation of mullite ceramics to Al_2O_3 conglomerates mechanically bonded with SiO_2 of an initial size equal to ~20 μm (Figure 4a). A detailed analysis of the primary oxide ceramics areas revealed the initiation of metallic liquid infiltration into the porous ceramics. During sintering at 1030 °C, the defragmentation process of primary SiO_2 precipitates was observed, as presented in Figure 4b. Since the liquid metal infiltrated the porous silica ceramic, the silicon atoms diffused into the $FeAl_3$ phase. The primary two-component matrix was transformed into the ternary $Al_{4.5}FeSi$ (τ_6) phase (Figure 4c). Additionally, the oxygen atoms diffused from SiO_2 ceramics and reacted with the most chemically active aluminum atoms taken from the metallic liquid. As a consequence, fine, spherical Al_2O_3 particles (with diameters less than 1 μm) in the iron–aluminum–silicon matrix were formed (Figure 4d). The mechanism observed for growth of new, fine Al_2O_3 aluminum oxide precipitates confirmed the correctness of the microstructure evolution model.

The SHS reaction provided a large amount of energy and subsequent immediate temperature increase. The temperature increase led to melting of the $FeAl_3$ matrix and its transformation to $Al_{4.5}FeSi$ ($\tau 6$). Such material behavior during the SHS process, where the external pressure during sintering was not provided, may lead to the formation of porous structures, as reported in [43]. Subsequent homogenization should accelerate the densification process and, as a consequence, increase the density of the material.

4.3. X-ray Phase Analysis of Sintered Material

The assumptions of the material model were subsequently confirmed by using in situ X-ray phase analysis. The diffractograms were recorded at the actual test temperature ranging from 25 to 900 °C (Figure 5). X-ray in situ analysis was performed by using a unique, high-temperature HTK attachment. It allowed for phase structure analysis as the function of time and temperature. In order to ensure the correctness of temperature selection, the sample was annealed for 1 h before each in situ X-ray diffraction measurement. The following phase transformations were observed:

- At temperatures up to 500 °C, the material mainly consisted of Al_2O_3, SiO_2, Fe_2Al_9 and Al_2FeSi phases. Such phase compositions indicated that the temperature was too low to initiate phase transformations.
- In the temperature range of 500 to 600 °C, the material mainly consisted of the Al_2O_3 phase and $Al_2Fe_3Si_3$ and Fe_2Al_9 phases, which were presumably indirect products of the phase transformations initiated.
- At the temperature of 700 °C, an intense peak of Al_2O_3 was observed. An occurrence of such a peak was associated with the formation of new, fine precipitates of this oxide.
- At temperatures higher than 700 °C, the sinter phase composition stabilized. It consisted of the $Al_{4.5}FeSi$ phase and new Al_2O_3. The homogenization at the temperature of 900 °C did not affect the phase structure of the material significantly because only one additional peak of the $Al_2Fe_3Si_3$ phase was observed.

- The Halder–Wagner (HW) method was used to determine crystallite size in the function of temperature (Figure 6). The size of Al_2O_3 crystallites increased with the temperature, and the average value of 392 Å was obtained after heating to 900 °C.

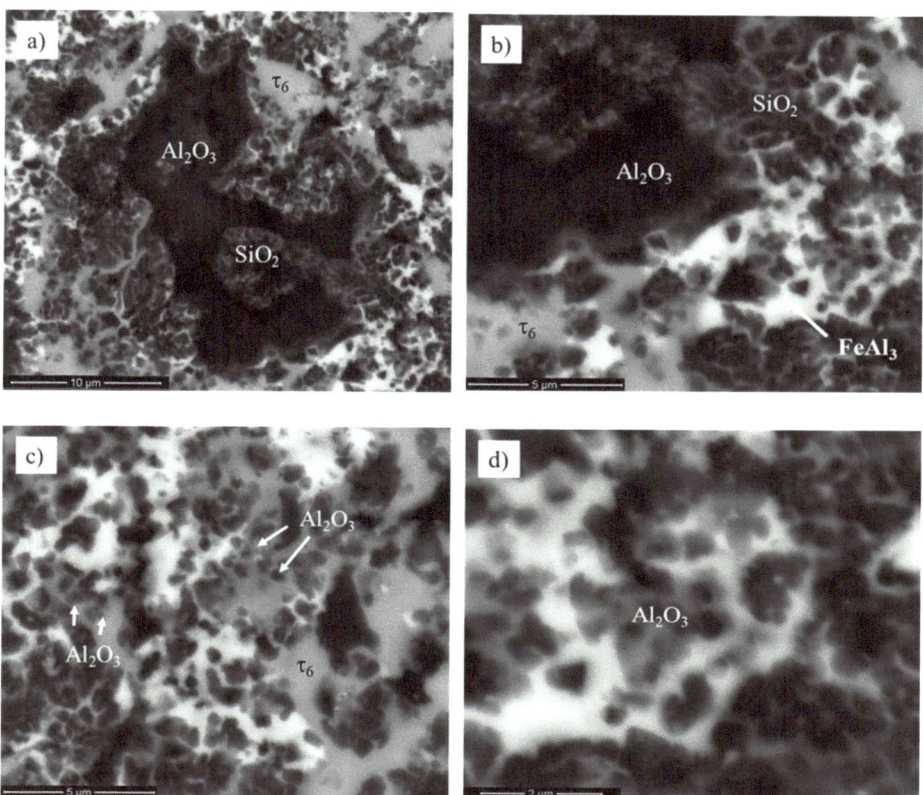

Figure 4. Mechanically bonded Al_2O_3/SiO_2 conglomerates (**a**); defragmentation of the silica ceramics through infiltration (**b**); growth of fine Al_2O_3 oxides in τ_6 matrix (**c**); structure of spherical Al_2O_3 particles within the matrix (**d**).

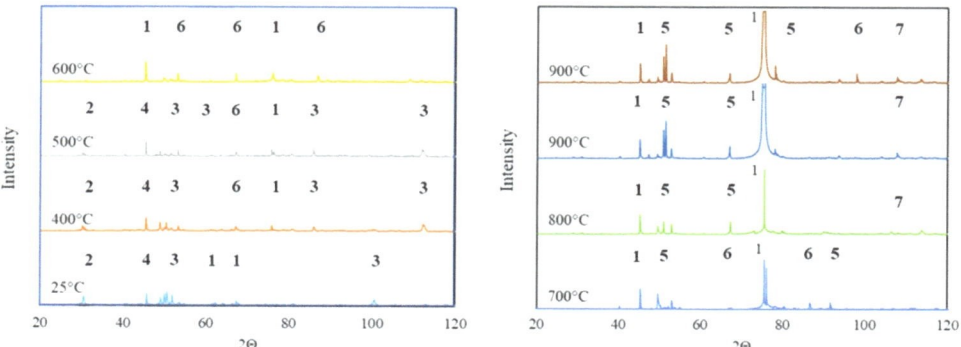

Figure 5. Phase structure evolution during heating from 25 to 900 °C, where 1—Al_2O_3, 2—SiO_2, 3—Fe_2Al_9, 4—Al_2FeSi, 5—$Al_{4.5}FeSi$, 6—$Al_2Fe_3Si_3$ and 7—$FeAl_3$.

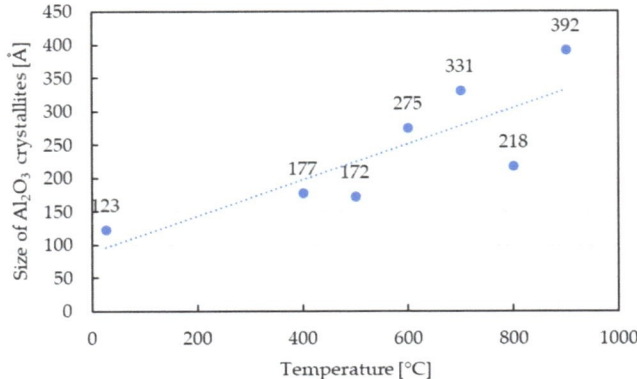

Figure 6. Evolution of Al_2O_3 crystallites size during heating.

5. Conclusions

The performed microstructural characterization and further analysis confirmed the correctness of the proposed model, aimed primarily at obtaining fine precipitates of Al_2O_3 oxide reinforcement in the intermetallic matrix of the composite. However, during the classical sintering process in the graphite dies using external pressure, the main problems were associated with the transition of silicon into the metallic liquid. This phenomenon led to an increase of the metallic liquid phase fluidity. Such material behavior combined with external pressure and caused loss of material during the sintering process. It was proposed, therefore, to perform the sintering process without external pressure. The improved sintering process allowed for successful formation of the composite material. Moreover, the occurrence of the assumed phase transformations was observed, in particular the defragmentation of SiO_2 oxide and the construction in its primary areas of fine, spherical Al_2O_3 oxide precipitations, arranged in a ternary intermetallic iron–aluminum–silicon matrix. Based on the studies performed showing the correctness of the proposed model, one can recommend it for further usage in manufacturing of IMC composites. It should be emphasized here that the application of external loads during the sintering process to prevent the outflow of metallic liquid from infiltrating ceramics is crucial and must be taken into account. It was concluded that a technology potentially capable of ensuring successful sintering of IMCs could be hot isostatic pressing (HIP).

Summing up the results of this work, one can state in particular:

- The proposed model of composite fabrication, based on the decomposition of mullite ceramics infiltrated with a metallic liquid, allows for obtainment of a composite material with an intermetallic matrix reinforced with Al_2O_3 ceramic particles derived from the defragmentation of primary alundum ceramics and newly formed fine-grained spherical ceramics resulting from SiO_2 decomposition.
- The parameters of the sintering process and subsequent homogenization are significantly affected by the participation of silicon, which significantly increases the fluidity of the metallic liquid formed during sintering and the final phase structure of the sintered matrix.
- The correctness of the assumed model must be supported by using proper sintering conditions that would lead to satisfactory cohesion between the matrix and reinforcement. Future promising sintering technology seems to be hot isostatic pressing that should guarantee fabrication of IMC composites with satisfactory performance.

Author Contributions: Conceptualization, M.K. and S.J.; methodology, M.K. and S.J.; formal analysis, Z.L.K.; writing—original draft preparation, M.K.; writing—review and editing, Z.L.K.; supervision, S.J. All authors have read and agreed to the published version of the manuscript.

Funding: This research received no external funding.

Conflicts of Interest: The authors declare no conflict of interest.

References

1. Kainer, K.U. *Basics of Metal Matrix Composites*; WileyVCH Verlag GmbH & Co. KGaA: Weinheim, Germany, 2006. [CrossRef]
2. Rosso, M. Ceramic and metal matrix composites: Route and properties. *J. Mater. Process. Technol.* **2006**, *175*, 364–375. [CrossRef]
3. Sritharan, T.; Murali, S.; Hing, P. Exothermic reactions in powder mixtures of Al, Fe and Si. *Mater. Lett.* **2001**, *51*, 455–460. [CrossRef]
4. Murali, S.; Sritharan, T.; Hing, P. Self-propagating high temperature synthesis of AlFeSi intermetallic compound. *Intermetallics* **2003**, *11*, 279–281. [CrossRef]
5. Gialanella, S.; Camana, G.; Kazior, J.; Lutterotti, L.; Pieczonka, T.; Molinari, A. Reaction sintering of Fe-Al-Si alloys. *J. Phase Equilib.* **2002**, *23*, 72. [CrossRef]
6. Gupta, S.P. Intermetallic compound formation in Fe-Al-Si ternary system: Part I. *Mater. Charact.* **2003**, *49*, 269–291. [CrossRef]
7. Maitra, T.; Gupta, S.P. Intermetallic compound formation in Fe-Al-Si ternary system: Part II. *Mater. Charact.* **2003**, *49*, 293–311. [CrossRef]
8. Novák, P.; Zelinková, M.; Serák, J.; Michalcová, A.; Novák, M.; Vojt, D. Oxidation resistance of SHS Fe–Al–Si alloys at 800°C in air. *Intermetallics* **2011**, *19*, 1306–1312. [CrossRef]
9. Novák, P.; Průša, F.; Knotek, V.; Šerák, J.; Vojtěch, D. Properties of intermetallic phases prepared by reactive sintering. In Proceedings of the 19th International Conference on Metallurgy and Materials 2010, Rožnov pod Radhoštěm, Czech Republic, 18–20 May 2010.
10. Sritharan, T.; Murali, S. Synthesis of ternary intermetallics by exothermic reaction. *J. Mater. Process. Technol.* **2001**, *113*, 469–473. [CrossRef]
11. Chen, K.; Liu, G.; Li, J. Combustion synthesis of refractory and hard materials: A review. *Int. J. Refract. Met Hard Mater.* **2013**, *39*, 90–102. [CrossRef]
12. Yadav, T.P.; Yadav, R.M.; Singh, D.P. Mechanical milling: A top down approach for the synthesis of nanomaterials and nanocomposites. *J. Nanosci. Nanotechnol.* **2012**, *2*, 22–48. [CrossRef]
13. Halldearn, R.D.; Xiao, P. Mechanisms of the aluminium-iron oxide thermite reaction. *Scr. Mater.* **1999**, *41*, 541–548. [CrossRef]
14. Nash, J.M.; Williams, J.C.; Breslin, M.C.; Daehn, G.S. Co-continuous composites for high-temperature applications. *Mat Sci Eng A* **2007**, *463*, 115–121. [CrossRef]
15. Han, C.Z.; Brown, I.W.M.; Zhang, D.L. Microstructure development and properties of alumina—Ti aluminide interpenetrating composites. *Curr. Appl. Phys.* **2006**, *6*, 444–447. [CrossRef]
16. Siqueira, J.R.R.; Simões, A.Z.; Stojanovic, B.D.; Paiva-Santos, C.O.; Santos, L.P.S.; Longo, E.; Varela, J.A. Influence of milling time on mechanically assisted synthesis of $Pb_{0.91}Ca_{0.1}TiO_3$ powders. *Cer. Inter.* **2007**, *33*, 937–941. [CrossRef]
17. Chen, W.; Xiao, H.; Fu, Z.; Fang, S.; Zhu, D. Reactive hot pressing and mechanical properties of TiAl 3-Ti3AlC2-Al2O3. *Mater. Des.* **2013**, *49*, 929–934. [CrossRef]
18. Inoue, M.; Nagao, H.; Suganuma, K.; Niihara, K. Fracture properties of Fe 40 at % Al matrix composites reinforced with ceramic particles and fibers. *Mat. Sci. Eng. A* **1998**, *258*, 298–305. [CrossRef]
19. Avraham, S.; Beyer, P.; Janssen, R.; Claussen, N.; Kaplan, W.D. Characterization of Al_2O_3–$(Al–Si)_3Ti$ composites. *J. Eur. Ceram. Soc.* **2006**, *26*, 2719–2726. [CrossRef]
20. Zhu, H.X.; Abbaschian, R. In-situ processing of NiAl–alumina composites by thermite reaction. *Mat Sci Eng A* **2000**, *282*, 1–7. [CrossRef]
21. Bauer, R.S.; Bachrach, R.Z.; Brillson, L.J. Au and Al interface reactions with SiO_2. *Appl. Phys. Lett.* **1980**, *1006*, 35–38. [CrossRef]
22. Liu, W.; Küster, U. Criteria for formation of interpenetrating oxide/metal-composites by immersing sacrificial Oxide preforms in molten metals. *Scr. Mater.* **1996**, *35*, 35–40. [CrossRef]
23. Wagner, F.; Garcia, D.E.; Krupp, A.; Claussen, N. Interpenetrating Al_2O_3-$TiAl_3$ alloys produced by reactive infiltration. *J. Eur. Ceram. Soc.* **1999**, *19*, 2449–2453. [CrossRef]

24. Fan, R.; Sun, K.; Wang, W.; Yi, X. Kinetics of thermite reaction in Al-Fe$_2$O$_3$ system. *Thermochim. Acta* **2006**, *440*, 129–131. [CrossRef]
25. Manfredi, D.; Pavese, M.; Biamino, S.; Fino, P.; Badini, C. Preparation and properties of NiAl(Si)/Al$_2$O$_3$ co-continuous composites obtained by reactive metal penetration. *Compos. Sci. Technol.* **2009**, *69*, 1777–1782. [CrossRef]
26. Schicker, S.; Garcia, D.E.; Bruhn, J.; Janssen, R.; Claussen, N. Reaction synthesized Al$_2$O$_3$-based Intermetallic composites. *Acta Mater.* **1998**, *46*, 2485–2492. [CrossRef]
27. Bhatt, J.; Balachander, N.; Shekher, S.; Karthikeyan, R.; Peshwe, D.R.; Murty, B.S. Synthesis of nanostructured Al-Mg-SiO$_2$ metal matrix composites using high-energy ball milling and spark plasma sintering. *J. Alloys Compd.* **2012**, *536*, S35–S40. [CrossRef]
28. Missiaen, J. Modelling of sintering: recent developments and perspectives. *Rev. Met. Paris* **2002**, *12*, 1009–1019. [CrossRef]
29. Exner, H.; Kraft, T. Review on computer simulations of sintering processes. *Powder Metall. World Congr.* **1998**, *2*, 278–283.
30. Pan, J. Modelling sintering at different length scales. *Int. Mater. Rev.* **2003**, *2*, 69–85. [CrossRef]
31. Parhami, F.; McMeeking, R. A network model for initial stage sintering. *Mech. Mater.* **1998**, *27*, 111–124. [CrossRef]
32. Martin, C.; Schneider, L.; Olmos, L.; Bouvard, D. Discrete element modeling of metallic powder sintering. *Scr. Mater.* **2006**, *55*, 425–428. [CrossRef]
33. Park, M.S.; Arroyave, R. Early stages of intermetallic compound formation and growth during lead-free soldering. *Acta Mater.* **2010**, *58*, 4900–4910. [CrossRef]
34. Liu, S.; Shin, Y.C. Simulation and experimental studies on microstructure evolution of resolidified dendritic TiCx in laser direct deposited Ti-TiC composite. *Mater. Des.* **2018**, *159*, 212–223. [CrossRef]
35. Roehling, J.D.; Perron, A.; Fattebert, J.-L.; Haxhimali, T.; Guss, G.; Li, T.T.; Bober, D.; Stokes, A.W.; Clarke, A.J.; Turchi, P.E.A.; et al. Rapid solidification in bulk Ti-Nb alloys by single-track laser melting. *JOM J. Miner. Met. Mater. Soc.* **2018**, *70*, 1589–1597. [CrossRef]
36. Keller, T.; Lindwall, G.; Ghosh, S. Application of finite element, phase-field, and CALPHAD-based methods to additive manufacturing of Ni-based superalloys. *Acta Mater.* **2017**, *139*, 244–253. [CrossRef]
37. Ferreira, A.F.; Paradela, K.G.; Junior, P.F.; Júnior, Z.A.; Garcia, A. Phase-field simulation of microsegregation and dendritic growth during solidification of hypoeutectic Al-Cu alloys. *Mater. Res.* **2017**, *20*, 423–429. [CrossRef]
38. Bhaskar, M.S. Quantitative phase field modelling of precipitate coarsening in Ni-Al-Mo alloys. *Comput. Mater. Sci.* **2018**, *146*, 102–111. [CrossRef]
39. Aristizabal, K.; Katzensteiner, A.; Bachmaier, A. Microstructural evolution during heating of CNT/Metal Matrix Composites processed by Severe Plastic Deformation. *Sci. Rep.* **2020**, *10*, 857. [CrossRef]
40. Długosz, P.; Darłak, P.; Purgert, R.M.; Sobczak, J.J. Technological aspects of the synthesis of metal matrix composites reinforced with cenospheres. *Prace Instytutu Odlewnictwa* **2011**, *51*, 35–44.
41. Raghavan, V. Al-Fe-Si (Aluminum-Iron-Silicon). *J. Phase Equilib.* **2002**, *23*, 362–366. [CrossRef]
42. Siemiaszko, D.; Jóźwiak, S.; Czarnecki, M.; Bojar, Z. Influence of temperature during pressure-assisted induction sintering (PAIS) on structure and properties of the Fe40Al intermetallic phase. *Intermetallics* **2013**, *41*, 16–21. [CrossRef]
43. Pocheć, E.; Jóźwiak, S.; Karczewski, K.; Bojar, Z. Fe–Al phase formation around SHS reactions under isothermal conditions. *J. Alloys Compd.* **2011**, *509*, 1124–1128. [CrossRef]

© 2020 by the authors. Licensee MDPI, Basel, Switzerland. This article is an open access article distributed under the terms and conditions of the Creative Commons Attribution (CC BY) license (http://creativecommons.org/licenses/by/4.0/).

MDPI AG
Grosspeteranlage 5
4052 Basel
Switzerland
Tel.: +41 61 683 77 34

Materials Editorial Office
E-mail: materials@mdpi.com
www.mdpi.com/journal/materials

Disclaimer/Publisher's Note: The title and front matter of this reprint are at the discretion of the Guest Editor. The publisher is not responsible for their content or any associated concerns. The statements, opinions and data contained in all individual articles are solely those of the individual Editor and contributors and not of MDPI. MDPI disclaims responsibility for any injury to people or property resulting from any ideas, methods, instructions or products referred to in the content.

www.ingramcontent.com/pod-product-compliance
Lightning Source LLC
LaVergne TN
LVHW072359090526
838202LV00019B/2584